NOW 2 kNOW!

Electro-Magnetic Fields

by T. G. D'Alberto

Pithy Professor Publishing Company

Brighton, CO

Published by

Pithy Professor Publishing Company, LLC
PO Box 33824
Northglenn, CO 80233

ISBN: 978-0-9882054-1-3

Library of Congress Control Number: 2012917641

Printed in the United States of America

About the Author

Dr. Tiffanie G. D'Alberto has a Ph.D. in Electrical & Computer Engineering from Cornell University and a B.S. and M.S. in Electrical Engineering from Virginia Polytechnic Institute & State University.

She served as a teaching assistant for EM Fields three times while at Cornell, and has engaged in numerous opportunities for tutoring, teaching, and mentoring throughout her career and schooling. She has worked for over a decade in the telecommunications and aerospace industries as a scientist, project manager, and supervisor.

In her spare time, Tiffanie enjoys oil painting, drawing, reading, sewing, and running. She's a huge fan of Star Trek, Renaissance Festivals, and animals.

Tiffanie lives in Colorado with her fiancé, Colin, and their many wonderful pets.

Dedication

*To my dearest Colin, who inspires me, encourages me, and
supports me. I could never thank you enough.*

*To my advisor, Dr. C. Pollock, who always found a way
to bring real life into the classroom.*

Acknowledgements

I always thank my family first: My parents for the foundation, the push, and the belief in me all along; My fiancé for his inspiration, encouragement, and unending support.

A huge thanks goes to my advisor, Dr. C. Pollock, who gave me the chance to help teach this subject at Cornell. He provided many insights to the material that solidified my understanding and helped me serve my students better.

Finally, I'd like to thank Amazon.com for their excellent publish-on-demand service that enables books such as these, and you, the reader, for making this investment in your future.

Table of Contents

Introduction   1

Welcome!   1

Layout   2

Part 1: Statics   3

Chapter 1: Maxwell's Equations for Statics   5

General Principles   5

Derivative Form of Maxwell's Equations   7

Integral Form of Maxwell's Equations   9

Chapter 2: Electro-Statics   13

Forms of Static Charge   13

Permittivity & Susceptibility   16

Coulomb's Law, Part 1   17

Chapter 3: Electric Potential   21

Electric Potential & The Electric Field   21

Electric Potential & Sources   23

Electric Dipoles   25

Chapter 4: Magneto-Statics — 27

Current, Current Density, & Conductance — 27
Permeability & Susceptiblity — 29
Biot-Savart's Law — 30
Magnetic Potential — 33
Magnetic Dipoles — 35

Chapter 5: Boundary Conditions — 37

Normal Boundary Conditions — 37
Tangential Boundary Conditions — 38
Image Theory — 41

Chapter 6: Circuit Measurements — 43

Resistance, Conductance, & Capacitance — 43
Flux & Inductance — 45
Power & Energy — 48

Chapter 7: Moments, Forces, & Torque — 51

Moments — 51
Coulomb Force — 53
Lorentz Force — 54
Torque — 56

Part 2: Dynamics   59

Chapter 8: Equations for Dynamics   61

Faraday's Law   61

Ampère's Law   64

Maxwell's Equations for Dynamics   66

Retarded Potentials   67

Chapter 9: Electro-Magnetic Waves   71

An Introduction to Waves   71

Maxwell's Equations for Waves   74

The Wave Equation   75

Power & Energy   76

Chapter 10: Polarization   79

Types of Polarization   79

Determining the Type of Polarization   81

Chapter 11: Boundaries at Normal Incidence   85

Normal Incidence at a Single Boundary   85

Normal Incidence at Multiple Boundaries   87

Propagation & Attenuation   89

Chapter 12: Boundaries at Oblique Incidence 93

Oblique Incidence 93

Perpendicular Polarization 94

Parallel Polarization 96

The Critical Angle 99

Brewster's Angle 100

Chapter 13: Transmission Lines 101

Transmission Line Characteristics 101

Reflections & Standing Waves 103

Input Impedance 105

Chapter 14: Transmission Line Analysis Tools 109

ABCD Analysis of Transmission Lines 109

Smith Chart Analysis of Transmission Lines 112

Transient Analysis of Transmission Lines 116

Appendices 121

Appendix A: Course Summary 122

Appendix B: Vector Calculus Review 125

Appendix C: Units & Constants 128

Index 129

Introduction

<u>Welcome!</u>

I've taught the recitation portion of this course several times at Cornell University. I've also taken it myself as an undergraduate overloaded with other coursework and a waitressing job. Through these experiences, I know that your chief concern is to figure out, in the shortest time possible, how to successfully do the homework and exam problems. So, let's get to it.

To be successful in this subject, you need five tools:

1. **Vector Calculus.** KNOW your Vector Calculus! After teaching my first class, I learned to always start day one of the semester with a review of this material. Don't worry, there's help in Appendix B.
2. **Sketches**. Do not attempt to perform integral calculations in this field without some sort of picture of the problem. Once you have a drawing, you'll be amazed at how quickly the right answer pops out. Without a drawing, you'd be amazed how much work you put into getting the wrong answers.
3. **Conceptual understanding.** Unfortunately, a description of the physical phenomena dealt with in this subject can be minimal in some texts that assume you are born with the knowledge or picked it up elsewhere. Most of us aren't that lucky, and having a working model in your head of what's going on is a needed first step.
4. **Examples.** Once you understand the equations and how they were derived, putting them into practice takes, well, practice. Studying examples is a great way to gain some painless practice before you try it on your own.
5. **The Big Picture.** In the many courses I've taken, I've learned that outlining the material helps organize the massive flow of information. Also, by summarizing the main ideas into an area big enough to see at a glance, you have a quick reference of the tools at your disposal to tackle the toughest problems.

Layout:

The layout of this text specifically addresses the five tools you need to succeed:

1. **Vector Calculus Review.** A handy review of coordinate systems; line, surface, and volume integrals; and operations in rectangular, cylindrical, and spherical coordinates is given in Appendix B. .
2. **Detailed Sketches.** Each major topic and integration example offer sketches that show you how to break a problem down into simple, solvable systems.
3. **Conceptual Descriptions.** Chapters 1 & 9 as well as sections throughout the text offer a discussion of the physical phenomena you'll be dealing with in this field.
4. **Numerous examples.** The text is filled with illustrative examples that are easy to follow.
5. **Course Summary.** Appendix A is an overall summary of the entire book that helps you visualize the big picture, organize the details, and view the tools you have acquired in the course.

In addition, the following visual markers will help you navigate the material...

Key terms defined for the first time are **bolded** and also found in the index.

Important equations are shown as:

> *important equations*

Illustrative graphics and additional notes are shown on the side to accompany the text.

Finally, examples are given as supplements to the text as well as for illustration:

> *Example* This is an example to illustrate a point or to give further definition. Skip it if you feel very comfortable with the material presented thus far.

Part 1: Statics

Chapter 1: Maxwell's Equations for Statics

General Principles:

The first thing to understand in electro-magnetics is the definition of the key players. The following table summarizes the main elements along with their symbols and units of measure (the first unit given is the most commonly used):

Element	Symbol	Units	
Electric Field	E	V/m	
Electric Flux Density	D	C/m²	A·s/m²
Permittivity	ε	F/m	C/V·m, A·s/V·m
Magnetic Field	H	A/m	C/m·s
Magnetic Flux Density	B	T	Wb/m², V·s/m²
Permeability	μ	H/m	Wb/A·m, V·s/A·m

The **electric field intensity** or **electric field, $\vec{E}$**, originates from charges, whether they are static, dynamic, localized, or distributed. A field is what allows a charge to act on physical phenomena that are not in direct contact with it. The charge can exert a force, impart energy, or give rise to magnetism through the action of the field. The electric field at any given point can be measured as the amount of electrical force per unit charge experienced by a test charge at that point.

The **electric flux density, $\vec{D}$**, is a measure of charge per unit area and gives an indication of the environment through which the electric field propagates. It is related to the electric field by the **constituent relation**:

$$\vec{D} = \varepsilon \vec{E}$$

where ε is the **permittivity** of the material or vacuum.

Permittivity can be thought of as the effect the material has on the propagation of the electric field. Material affects include refraction, retardation, attenuation, anisotropic effects, and non-linear effects. The materials that we will consider in this text will be linear and isotropic which means that ε will be a constant.

Similar to the electric field, the **magnetic field intensity** or **magnetic field**, $\vec{H}$, also originates from charge and gives rise to the magnetism force of the charge through space. The **magnetic flux density**, $\vec{B}$, is a measure of the magnetic flux, or flow of magnetic field, from the charge. Whereas $\vec{D}$ is proportional to charge, $\vec{B}$ is proportional to potential.

The magnetic field and magnetic flux density are related by the **constituent relation**:

$$\vec{B} = \mu\vec{H}$$

where μ is the **permeability** of the material or vacuum through which the fields propagate. Like permittivity, permeability is the effect of the material on the magnetic field. It is usually a constant except in the case of ferromagnetic materials.

Because both the electric field and magnetic field originate from charge, one would expect that these two fields are closely related. In actuality, there are. Their propagations and interactions are closely coupled, whether we're dealing with static charges or moving charges.

Fortunately, a scientist named James Clerk Maxwell defined these relationships in 1873 pulling together the work of a number of scientists into a unified theory of electromagnetism. The theory is referred to as **Maxwell's equations**.

Derivative Form of Maxwell's Equations:

The **derivative form (or point form) of** Maxwell's **equations** for static electro-magnetics is given as follows:

$$\nabla \times \vec{E} = 0 \qquad\qquad \nabla \times \vec{H} = \vec{J}$$

$$\nabla \cdot \vec{D} = \rho_v \qquad\qquad \nabla \cdot \vec{B} = 0$$

Kirchhoff's law, $\nabla \times \vec{E} = 0$, states that the curl of the electric field is zero. This means that the circulation of the electric field is zero everywhere, i.e. the electric field is uniform and does not circulate when dealing with static charges. We'll see that this is not true of dynamic charges.

Ampère's law, $\nabla \times \vec{H} = \vec{J}$, however, states that a current gives rise to circulation of the magnetic field and vice versa. In particular, the circulation of $\vec{H}$ is equal to **current density**, $\vec{J}$, which has units of A/m². To determine the direction of $\vec{J}$ and $\vec{H}$, you use the right hand rule: using your right hand, curl your fingers in the direction of the circulation of $\vec{H}$ and your thumb points in the direction of the induced current.

Gauss's laws for electro-statics and magneto-statics deal with the divergence of the fields. The first law, $\nabla \cdot \vec{D} = \rho_v$, states that a **volume charge**, ρ_v, having units of C/m³, gives rise to a divergent electric flux. This is indicative of how the static electric field begins on a positive charge and ends on a negative charge.

The second law, $\nabla \cdot \vec{B} = 0$, states that the magnetic field does not diverge. In other words, whatever magnetic field enters a volume is the same amount of magnetic field that leaves a volume. This is indicative of the fact that magnetic fields start and end on themselves in a loop.

To summarize, the electric fields diverge, but do not circulate; the magnetic fields circulate, but do not diverge.

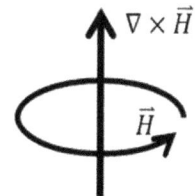

Circulation of magnetic field

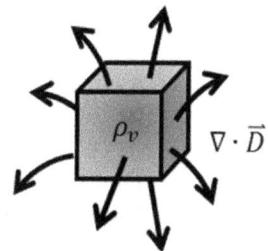

Divergence of electric field from a volume charge

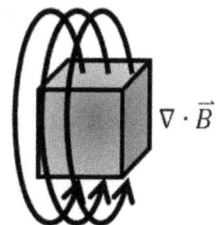

Divergence of magnetic field

Example

Find the volume charge associated with the following electric flux density:

$$\vec{D} = 2xyz\,\hat{x} + 4x^3y\,\hat{z} \quad nC/m^2$$

$$\rho_v = \nabla \cdot \vec{D}$$

$$= \frac{\partial}{\partial x}(2xyz) + \frac{\partial}{\partial y}(0) + \frac{\partial}{\partial z}(4x^3y)$$

$$= 2yz \quad nC/m^2$$

Example

Find the volume charge if:

$$\vec{D} = 2r^2 \sin\varphi\,\hat{r} + r^3 \sin\theta \sin\varphi\,\hat{\varphi} \quad nC/m^2$$

$$\rho_v = \nabla \cdot \vec{D}$$

$$= \frac{1}{r^2}\frac{\partial}{\partial r}(2r^4 \sin\varphi) + \frac{1}{r \sin\theta}\frac{\partial}{\partial\varphi}(r^3 \sin\theta \sin\varphi)$$

$$= 8\sin\varphi\, r + r^2 \cos\varphi \quad nC/m^2$$

Example

Find the current density in cylindrical coordinates if the magnetic field is:

$$\vec{H} = 2r^2\hat{r} + r\hat{\theta} \quad A/m$$

$\nabla \times \vec{H} = \vec{J}$ only the $\hat{z}$ terms are needed

$$\vec{J} = \left(\frac{1}{r}\frac{\partial H_z}{\partial\theta} - \frac{\partial H_\theta}{\partial z}\right)\hat{r} + \left(\frac{\partial H_r}{\partial z} - \frac{\partial H_z}{\partial r}\right)\hat{\theta}$$
$$+ \frac{1}{r}\left(\frac{\partial(rH_\theta)}{\partial r} - \frac{\partial H_r}{\partial\theta}\right)\hat{z}$$

$$= (0-0)\hat{r} + (0-0)\hat{\theta} + \frac{1}{r}\left(\frac{\partial(r^2)}{\partial r} - 0\right)\hat{z}$$

$$= 2\hat{z} \quad A/m$$

Integral Form of Maxwell's Equations:

The **integral form of Maxwell's equations** for static electro-magnetics is given as follows:

$$\oint \vec{E} \cdot d\vec{L} = 0 \qquad\qquad \oint \vec{H} \cdot dL = I$$

$$\oiint \vec{D} \cdot d\vec{S} = Q \qquad\qquad \oiint \vec{B} \cdot d\vec{S} = 0$$

Kirchhoff's law, $\oint \vec{E} \cdot d\vec{L} = 0$, states that the voltage (electric field times length gives units of volts) on any closed loop sums to zero. In other words, if you were to take a voltmeter, place one probe on a random point, take the other probe and place it on top of the first, your voltage reading would be zero. It doesn't matter how you construct your loop, either. If you take the second probe and whirl it around your head five times before closing the loop, your reading will still be 0 V.

$$\oint \vec{E} \cdot d\vec{L}$$

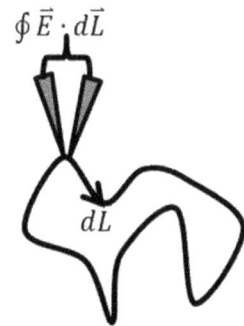

Voltage measurement on a closed loop

Ampère's law, $\oint \vec{H} \cdot d\vec{L} = I$, however, states that a current, I, gives rise to a circulation of the magnetic field (and vice versa). If you draw a loop that allows a path of current to pierce through, you will measure a magnetic field.

Gauss's laws for electro-statics and magneto-statics show how the fields propagate through a closed surface – think of a sphere or a cube. The first law, $\oiint \vec{D} \cdot d\vec{S} = Q$, states that an electric field is generated from charge, and if one encapsulates that charge within a closed surface, the electric field will transmit past that surface to end on a balancing equivalent charge. If the net charge within the surface is zero, or if there is no charge, the electric field will be contained, or will be equal to zero.

$$\oiint \vec{D} \cdot d\vec{S}$$

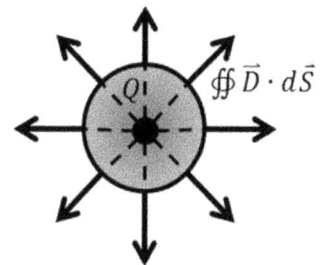

Propagation of electric field through a closed volume

The second law, $\oiint \vec{B} \cdot d\vec{S} = 0$, means that the sum of the magnetic field that leaves a closed surface is the sum of the magnetic field that enters the closed surface. Again, this is because magnetic fields occur in loops.

Example

Find the electric field associated with a single charge of 4 nC.

When using integrals, draw a picture.

We're going to set up a Gaussian surface to perform the integral. The surface will be closed and contain the charge in a geometry that makes sense. In our case, a sphere would meet the need, and as we'll see, the electric field lines will be perpendicular to the surface everywhere.

Closed surface surrounding a charge, Q

We will use the equation $\oiint \vec{D} \cdot d\vec{S} = Q$. In spherical coordinates:

$$d\vec{S} = r^2 \sin \theta \, d\theta \, d\varphi \, \hat{r}$$

$$\oiint (D_r \hat{r} + D_\theta \hat{\theta} + D_\varphi \hat{\varphi}) \cdot r^2 \sin \theta \, d\theta \, d\varphi \, \hat{r}$$

$$= \int_0^{2\pi} \int_0^{\pi} D_r \, r^2 \sin \theta \, d\theta \, d\varphi$$

$$= D_r 2\pi r^2 (\cos 0 - \cos \pi) = 4\pi r^2 D_r = Q$$

$$\vec{D} = \frac{Q}{4\pi r^2} \hat{r} = \frac{1}{\pi r^2} \hat{r} \ \ nC/m^2$$

$$\vec{E} = \frac{\vec{D}}{\varepsilon} = \frac{Q}{4\varepsilon \pi r^2} \hat{r} = \frac{1}{\pi \varepsilon r^2} \hat{r} \ \ nV/m$$

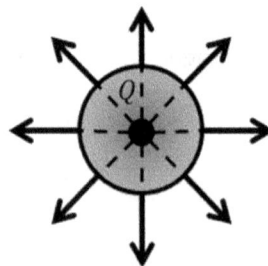

Example

Find the magnetic field due to a straight wire carrying a current of 5 A.

We can choose our coordinate system and the wire orientation as we wish. We'll assume the current points in the $+\hat{z}$ direction. To use Ampère's law, we have to draw a path around the current, so we'll draw a circle of radius r around the wire. Using the right hand rule, the right thumb points in the direction of the current which means the loop goes in the $+\hat{\theta}$ direction. With this information, we can say:

Closed loop surrounding a current, I.

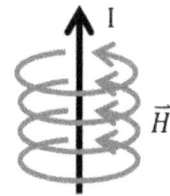

$$d\vec{L} = rd\theta\ \hat{\theta}$$
(there are no $\hat{r}$ and $\hat{z}$ components of $d\vec{L}$)

$$\oint \vec{H} \cdot d\vec{l} = \int H_r\hat{r} + H_\theta\hat{\theta} + H_z\hat{z} \cdot d\theta\ \hat{\theta}$$

$$= \int_0^{2\pi} H_\theta rd\theta = 2\pi rH_\theta = I$$

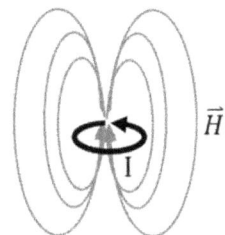

Resulting field.

$$\vec{H} = \frac{I}{2\pi r}\hat{\theta} = \frac{5}{2\pi r}\hat{\theta}\ \text{A/m (cylindrical coordinates)}$$

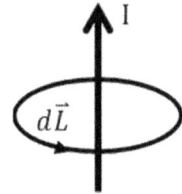

Example

Find the magnetic field due to a 5 A current loop.

Here we draw the path in spherical coordinates such that the current pierces the path once:

Current loop.

$$d\vec{L} = rd\theta\ \hat{\theta};$$

$$\oint \vec{H} \cdot d\vec{l} = \int H_r\hat{r} + H_\varphi\hat{\varphi} + H_\theta\hat{\theta} \cdot d\theta\ \hat{\theta}$$

$$= \int_0^{2\pi} H_\theta rd\theta = 2\pi rH_\theta = I$$

$$\vec{H} = \frac{I}{2\pi r}\hat{\theta} = \frac{5}{2\pi r}\hat{\theta}\ \text{A/m (spherical coordinates)}$$

$$\vec{H} = \frac{I}{2\pi\sqrt{r^2+z^2}}[\cos\theta\ \hat{r} - \sin\theta\ \hat{z}]\ \text{A/m (cylindrical)}$$

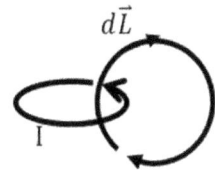

Resulting field.

Chapter 2: Electro-Statics

Forms of Static Charge:

The following table lists the basic types of charge that are dealt with in electro-magnetics:

Element	Symbol	Units
Point Charge	q	C
Total Charge	Q	C
Line Charge Density	ρ_L	C/m
Surface Charge Density	ρ_S	C/m^2
Volume Charge Density	ρ_V	C/m^3
Current	I	A = C/s
Current Density	J	A/m^2

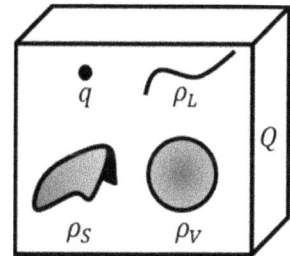

Different kinds of charges dealt with in electro-magnetics including Q, the total charge

The **point charge**, q, is a highly localized ion, proton, or other charged particle. When the charged particle is not localized, we can talk about charge densities. A **line charge density**, ρ_L, is the charge present along a path (which needs not be straight). A **surface charge density**, ρ_S, is charge distributed along a surface, and a **volume charge density**, ρ_V, is charge distributed in a volume. Each of these quantities can be expressed as:

$$\rho_L = \frac{dq}{dL}; \quad \rho_S = \frac{dq}{dS}; \quad \rho_V = \frac{dq}{dV}.$$

The sum of all charges in a volume is captured by the quantity, Q, or **total charge**, given by:

$$Q = \sum q_i + \int \rho_L dL + \int \rho_S dS + \int \rho_V dv.$$

Current, I, is defined as charge that moves (with constant velocity in the case of electro-statics) and the **current density**, $\vec{J}$, is a vector quantity whose magnitude is the amount of current flowing through a cross-sectional area:

$$I = \int \vec{J} \cdot d\vec{S}.$$

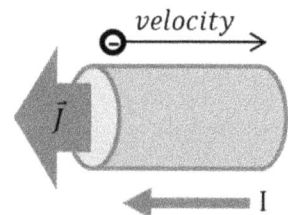

Current is the movement of charge and points away from the flow of electrons. Current density is the current flux through a cross-sectional area.

13

Example

Find the charge carried on a tube with length 1 m, radius 2 cm, and surface charge density 4 C/m².

To find the charge, Q, we'll use cylindrical coordinates. The normal to the charged surface, $\widehat{n_s}$, will point in the $+\hat{r}$ direction giving us:

$$d\vec{S} = r\, dz\, d\theta\, \hat{r}$$

$$Q = \int_0^1 \int_0^{2\pi} \rho_s\, r\,d\theta\, dz = 2\pi r \rho_s = 160\pi \ \text{mC}$$

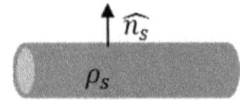

Surface charge density on a tubular surface.

Example

Find the charge carried on a disk with radius 1 cm and surface charge density 9 C/m².

We'll put the disk in the x-y plane with the surface normal pointing in the $+\hat{z}$ direction giving us:

$$d\vec{S} = r\, dr\, d\theta\, \hat{z}$$

$$Q = \int_0^{.01} \int_0^{2\pi} \rho_s\, r\,d\theta\, dr = \pi r^2 \rho_s = 0.9\pi \ \text{mC}$$

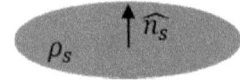

Surface charge density on the surface of a disk.

Example

Find the current carried in a wire that is 4 mm thick with 6 A/m² flowing through it.

We use cylindrical coordinates again, but in this case, electrons are flowing through the surface area of the wire. We can then write:

$$d\vec{S} = r\, dr\, d\theta\, \hat{z}; \text{ and } \vec{J} = J_z \hat{z} = 6\hat{z} \ \text{A/m}^2$$

It doesn't matter that we've turned the axes on their side as long as we're consistent.

Current being carried through a wire.

$$I = \int_0^{.02} \int_0^{2\pi} J_z\, r\,d\theta\, dr = \pi r^2 J_z = 2.4\pi \ \text{mA}$$

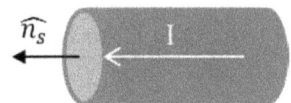

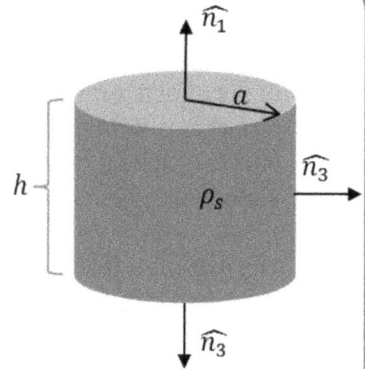

Find the electric field due to a cylinder carrying a surface charge, ρ_s.

We'll start with a cylinder of height, h, and radius, a. The Gaussian surface we'll draw around it is just another cylinder. We then have three surfaces:

$$d\vec{S_1} = rdr\,d\theta\,\hat{z}; \; d\vec{S_2} = rdr\,dz\,\hat{r}; \; d\vec{S_3} = -rdr\,d\theta\,\hat{z};$$

$$\oiint \vec{D} \cdot d\vec{S} = \int_0^{2\pi} \int_0^a D_{z-top} rdrd\theta$$

$$+ \int_0^{2\pi} \int_0^h D_r rdrdz - \int_0^{2\pi} \int_0^a D_{z-bottom} rdrd\theta$$

$$= D_{z-top}\pi a^2 + D_r 2\pi rh - D_{z-bottom}\pi a^2$$

Cylindrical surface charge.

If we let $h \to 0$, we have a circular sheet of charge. On the top and bottom we have:

$$Q = \int_0^a \int_0^{2\pi} \rho_s\, rd\theta\, dr = \pi a^2 \rho_s = D_z \pi a^2$$
$$Q = \int_0^a \int_0^{2\pi} -\rho_s\, rd\theta\, dr = -\pi a^2 \rho_s = -D_z \pi a^2$$

$$\vec{D} = \rho_s\hat{z};$$

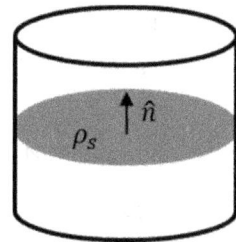

Geometry for infinite sheet charge.

You can imagine making a infinitely large giving:

$$\vec{D} = \rho_s\hat{z}; \quad \vec{E} = \frac{\rho_s}{\varepsilon}\hat{z} \quad \text{Infinite Sheet Charge}$$

Now consider the case where $a \to 0$. In that case, we are talking about a line charge where $\rho_L = 2\pi a\rho_s$

$$Q = \int_0^h \int_0^{2\pi} \rho_s\, rd\theta\, dr = 2\pi ah\rho_s = h\rho_L = D_r 2\pi rh$$

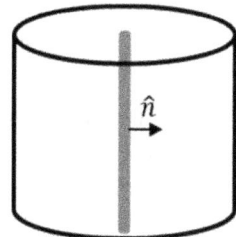

$$\vec{D} = \frac{\rho_L}{2\pi r}\hat{r}; \quad \vec{E} = \frac{\rho_L}{2\pi\varepsilon r}\hat{r} \quad \text{Infinite Line Charge}$$

Geometry for infinite line charge.

15

Permittivity & Susceptibility:

As described in Chapter 1, **permittivity**, ε, is an expression of the effects a material has on the electric field which propagates through it. Permittivity covers a variety of affects including refraction, retardation, attenuation, anisotropic effects, and non-linear effects.

The materials that we will consider in this text will be described by a constant permittivity. It is convenient to describe the **permittivity of free space**, ε_0, by which all other materials are compared. The value of this quantity is:

$$\varepsilon_0 = 8.854 \times 10^{-12} \text{ F/m.}$$

For a vacuum, air, and most conductors, the free space permittivity adequately describes the material. However, dielectrics behave differently. In a dielectric, an applied electric field can shift the polarity of individual atoms creating a polarization field, $\vec{P}$. This field is given by:

$$\vec{P} = \varepsilon_0 \chi_e \vec{E}$$

where χ_e is the **electric susceptibility**. The susceptibility is a measure of how easily the dielectric responds to the electric field and the strength of the resulting polarization field. The electric flux density, $\vec{D}$, including the polarization field becomes:

$$\vec{D} = \varepsilon_0 \vec{E} + \vec{P} = \varepsilon_0 (1 + \chi_e) \vec{E} \equiv \varepsilon \vec{E}.$$

By fulfilling the constituent relation $\vec{D} = \varepsilon \vec{E}$, we get that $\varepsilon = \varepsilon_0 (1 + \chi_e)$. We can assign the quantity $(1 + \chi_e)$ to a new term called the **relative permittivity**, ε_r. The expression for permittivity now becomes:

$$\varepsilon = \varepsilon_0 (1 + \chi_e) = \varepsilon_0 \varepsilon_r.$$

In a conductor, an applied electric field can dislodge electrons from atoms thus creating a distinct negative and positive charged area.

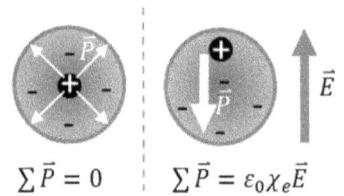

$\sum \vec{P} = 0$ $\sum \vec{P} = \varepsilon_0 \chi_e \vec{E}$

In a dielectric, an applied electric field causes the electron cloud to shift but not separate from the nucleus. The result is a polarization field.

The cumulative effect of an applied field to a dielectric is an overall shift in charge rather than a separation of charge.

Coulomb's Law, Part 1:

The first part of **Coulomb's law** states that the electric field experienced at any point, P, from a charge, q, is equal to:

$$\vec{E} = \frac{q}{4\pi\varepsilon r^2}\hat{r}$$

where ε is the permittivity of the material at the observation point, r is the distance from q to P, and $\hat{r}$ is the unit vector pointing from q to P.

Sometimes it is convenient to write this equation in an alternate form that references a global coordinate system. In this case, r is replaced by $r_o = |\vec{r_p} - \vec{r_s}|$ - the magnitude of the difference between the vector pointing from the origin to the observation point, $\vec{r_p}$, and the vector from the origin to the location of q, $\vec{r_s}$. The unit vector then becomes $(\vec{r_p} - \vec{r_s})/|\vec{r_p} - \vec{r_s}|$ to give:

$$\vec{E} = \frac{q}{4\pi\varepsilon|\vec{r_p}-\vec{r_s}|^2}(\vec{r_p} - \vec{r_s}) = \frac{q\hat{r_o}}{4\pi\varepsilon r_o^2}.$$

Similar to the electric field induced by a point charge, we can also write expressions for the field experienced from distributed charges. Here we integrate over the source which is indicated with a prime:

$$\vec{E} = \frac{1}{4\pi\varepsilon}\int\frac{\rho_L}{r_o^2}\hat{r_o}\,dL' + \frac{1}{4\pi\varepsilon}\int\frac{\rho_S}{r_o^2}\hat{r_o}\,dS' + \frac{1}{4\pi\varepsilon}\int\frac{\rho_V}{r_o^2}\hat{r_o}\,dv'.$$

Note that in the above equation we have summed the fields from multiple sources. We can do a summation because of the principle of **linear superposition** which states that electric fields add together. We can similarly treat the field from multiple point sources as:

$$\vec{E} = \sum\frac{q_i}{4\pi\varepsilon r^2}\hat{r} = \sum\frac{q_i(\vec{r_p}-\vec{r_i})}{4\pi\varepsilon|\vec{r_p}-\vec{r_i}|^2}.$$

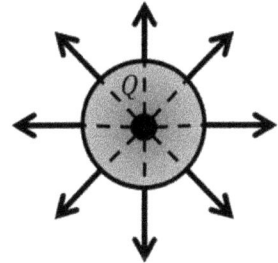

Coulomb's law states that the electric field starts from an electrical source. It is proportional to the magnitude of the charge and falls off as $1/r^2$ away from the charge.

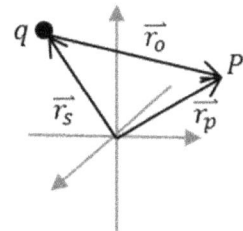

Vector representation of Coulomb's law.

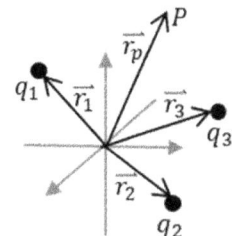

Vector representation of linear superposition.

Find the electric field from an infinite sheet of charge.

I'm not going to lie to you, Coulomb's law is confusing and hard to use at first. We start with:

$$\vec{E} = \frac{1}{4\pi\varepsilon} \int \frac{\rho_s}{r_o{}^2} \hat{r}_o \, dS'.$$

We pick a convenient point, P, to analyze that is on the z-axis. We will be integrating with respect to the _source_, so we want r_o in terms of the source variables:

$$r_o{}^2 = r_s{}^2 + z^2;$$
$$\hat{r}_o = \cos\alpha\,\hat{r} + \sin\alpha\,\hat{z} = \frac{r_s}{\sqrt{r_s{}^2+z^2}}\hat{r} + \frac{z}{\sqrt{r_s{}^2+z^2}}\hat{z};$$

Geometry for Coulomb's law evaluation of the infinite sheet of charge.

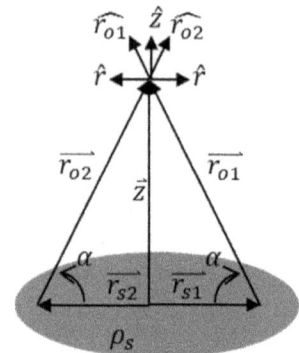

From the drawing, we can picture contributions to the field at point P coming from a circle of constant radius on the source. In that case, all of the $\hat{r}$ terms will cancel each other, and only the $\hat{z}$ terms will constructively add. As a result, we have

$$\hat{r}_o = \frac{z}{\sqrt{r_s{}^2+z^2}}\hat{z}$$

The integration variable, dS', is:

$$dS' = r_s\partial r_s\partial\theta;$$

Now we have:

$$\vec{E} = \frac{\rho_s}{4\pi\varepsilon} \int_0^{2\pi}\int_0^\infty \frac{z\,r_s\partial r_s\partial\theta}{\sqrt{r_s{}^2+z^2}^{\,3}}\hat{z}$$

$$= \frac{-2\pi\rho_s}{4\pi\varepsilon}\left(\frac{2}{2}\right)\hat{z}\left(\frac{z}{\sqrt{r_s{}^2+z^2}}\right)\Big|_0^\infty$$

$$= \frac{-\rho_s}{2\varepsilon}\hat{z}(0-1) = \frac{\rho_s}{2\varepsilon}\hat{z} \text{ V/m}$$

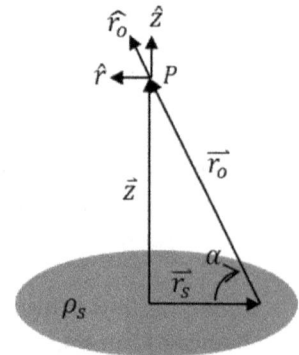

For the infinite sheet of charge, the $\hat{r}$ terms cancel.

18

Example

Find the electric field from an infinite line charge.

We start with:

$$\vec{E} = \frac{1}{4\pi\varepsilon} \int \frac{\rho_L}{r_o{}^2} \hat{r}_o \, dL'.$$

We pick a convenient point, P, to analyze that is on the r-axis. Once again, we will be integrating with respect to the source, so we want r_o in terms of the source variables. Once again we'll also define an angle, α, between the vectors $\vec{r_s}$ and $\vec{r_o}$.

$$r_o{}^2 = r_s{}^2 + r^2;$$
$$\hat{r}_o = \sin\alpha\,\hat{r} - \cos\alpha\,\hat{z} = \frac{r}{\sqrt{r_s{}^2 + r^2}}\hat{r} - \frac{r_s}{\sqrt{r_s{}^2 + r^2}}\hat{z};$$

Symmetry allows us to get rid of the $\hat{z}$ terms since only the $\hat{r}$ terms will constructively add. As a result, we have

$$\hat{r}_o = \frac{r}{\sqrt{r_s{}^2 + z^2}}\hat{r}$$

The integration variable, dL', is with respect to the source, so it becomes:

$$dL' = \partial r_s;$$

Now we have:

$$\vec{E} = \frac{\rho_L}{4\pi\varepsilon}\int_{-\infty}^{\infty} \frac{r\,\partial r_s}{\sqrt{r_s{}^2 + r^2}^3}\hat{r}$$

$$\int \frac{a\,du}{\sqrt{a^2 + u^2}^3} = \frac{u}{a^2\sqrt{a^2 + u^2}}$$

$$= \frac{\rho_L}{4\pi\varepsilon}\hat{r}\left(\frac{r\,r_s}{r^2\sqrt{r_s{}^2 + r^2}}\right)\Big|_{-\infty}^{\infty}$$

$$= \frac{\rho_L}{4\pi\varepsilon r}\hat{r}\left(\frac{r_s}{\sqrt{r_s{}^2 + r^2}}\right)\Big|_{-\infty}^{\infty} = \frac{\rho_L}{4\pi\varepsilon r}\hat{r}(1+1) = \frac{\rho_L}{2\pi\varepsilon r}\hat{r}\;\text{V/m}$$

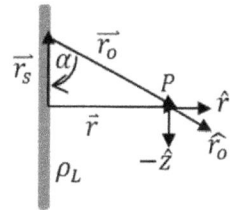

Geometry for Coulomb's law evaluation of the infinite line charge.

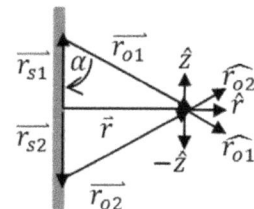

For the infinite line charge, the $\hat{z}$ terms cancel.

Chapter 3: Electric Potential

Electric Potential & The Electric Field:

Electric potential is the energy required to move a charge against an electric field. The electric potential is also commonly referred to as **voltage** and is the quantity that is measured with a voltmeter on a circuit.

As noted in Chapter 1, the electric potential is simply the electric field times the distance over which the field is measured. It has units of volts (V). In the most general sense, to measure the potential between points a and b, we use the following equation:

$$V_{ab} = V_b - V_a = -\int_a^b \vec{E} \cdot d\vec{L}.$$

It does not matter what path you take to measure the voltage, you will get the same answer every time. So, use the simplest path possible. Again, if you're measuring potential on a closed loop, $V = 0$ as per Maxwell's equations.

The voltage and electric field are also related through another equation. Considering only a differential potential element we get:

$$d(V) = -\vec{E} \cdot d\vec{L}.$$

We can further express this by writing out terms and noting that $dx = \hat{x} \cdot d\vec{L}, dy = \hat{y} \cdot d\vec{L},$ and $dz = \hat{z} \cdot d\vec{L}$:

$$d(V) = \left[\frac{\partial V}{\partial x}\hat{x} + \frac{\partial V}{\partial y}\hat{y} + \frac{\partial V}{\partial z}\hat{z}\right] \cdot d\vec{L} = \nabla V \cdot d\vec{L} = -\vec{E} \cdot d\vec{L}.$$

Therefore, we get:

$$\vec{E} = -\nabla V.$$

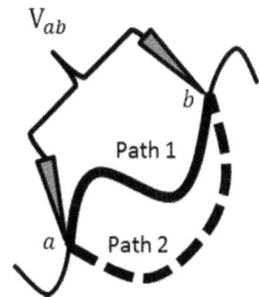

V_{ab}

Path 1

Path 2

Electric Potential between two points is measured via a line integral along a path, any path, connecting the two points.

Example

Find the potential difference between points $a(1,0,0)$ and $b(4,0,0)$ in an electric field of $\vec{E} = \frac{4}{x^2}\hat{x}$ V/m using three different paths.

The first path we'll choose is the direct one:

$$d\vec{L} = \partial x \hat{x};$$

$$V = -\int_1^4 \frac{4}{x^2}\hat{x} \cdot \partial x \hat{x} = -\int_1^4 \frac{4}{x^2}\partial x$$

$$= \frac{4}{x}\Big|_1^4 = 1 - 4 = -3 \text{ V}$$

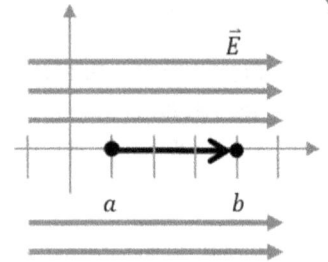

Diagram of problem with direct path.

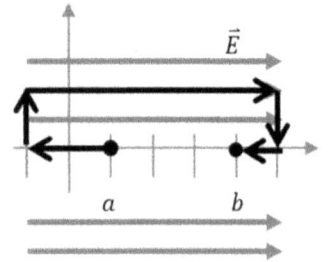

The second path will be to go backward to (-1,0,0), then up to (-1,2,0), then over to (5,2,0), down to (5,0,0), then finally to point b:

$$d\vec{L_1} = -\partial x \hat{x}; \; d\vec{L_2} = \partial y \hat{y}; \; d\vec{L_3} = \partial x \hat{x};$$
$$d\vec{L_4} = -\partial y \hat{y}; \; d\vec{L_5} = -\partial x \hat{x};$$

$$V = -[\int_{-1}^1 \frac{-4}{x^2}\hat{x} \cdot \partial x \hat{x} + \int_0^2 \frac{4}{x^2}\hat{x} \cdot \partial y \hat{y}$$
$$+ \int_{-1}^5 \frac{4}{x^2}\hat{x} \cdot \partial x \hat{x} + \int_0^2 \frac{-4}{x^2}\hat{x} \cdot \partial y \hat{y} + \int_4^5 \frac{-4}{x^2}\hat{x} \cdot \partial x \hat{x}$$

Diagram of problem with indirect path.

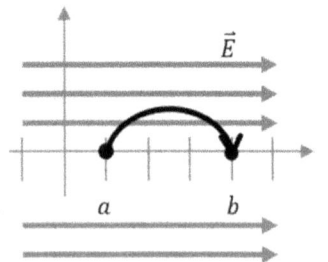

Luckily, all of the $\hat{x} \cdot \hat{y}$ terms are 0 leaving:

$$V = \frac{-4}{x}\Big|_{-1}^1 + \frac{4}{x}\Big|_{-1}^5 + \frac{-4}{x}\Big|_4^5$$

$$= -4 - 4 + \frac{4}{5} + 4 - \frac{4}{5} + 1 = -3 \text{ V}$$

Finally, we choose a circular path:

$$d\vec{L} = r\partial\theta\hat{\theta}; \; \vec{E} = \frac{4}{r^2\cos^2\theta}\left[\cos\theta\,\hat{r} - \sin\theta\,\hat{\theta}\right]$$

$$V = -\int_0^\pi \frac{-4\,r\,\sin\theta\partial\theta}{r^2\cos^2\theta} = \frac{-4}{r\cos\theta}\Big|_{r=4,\theta=0}^{r=-1,\theta=\pi}$$

Diagram of problem with circular path. Note that my convention is to integrate from 0 to π with a located at $(-1,\pi,0)$. Or, you can integrate from $(1,0,0)$ to $(4,0,0)$ in cylindrical coordinates.

$$= -4 + 1 = -3 \text{ V}$$

Electric Potential & Sources:

There are two other ways to find the electric potential and they involve using sources rather than the electric field. These methods are convenient when you have information about the sources and don't want to calculate the electric field before determining the resulting potential.

As stated in the prior section, you can measure the potential difference between any two points. One can also measure the potential between a reference point and point r away from the origin. A handy reference point that is normally used is $a = \infty$. The potential infinitely far away from the system is assumed to be zero (since the electric field falls off rapidly with r as per Coulomb's law). Therefore, we can rewrite the potential equation as:

$$V_r - 0 = V = -\int_{\infty}^{r} \vec{E} \cdot d\vec{L}.$$

Substituting Coulomb's equation and $d\vec{L} = dr\,\hat{r}$ gives:

$$V = -\int_{\infty}^{r} \frac{q}{4\pi\varepsilon r^2}\hat{r} \cdot \hat{r}\, dr = \frac{q}{4\pi\varepsilon r} = \frac{q}{4\pi\varepsilon|\vec{r_p}-\vec{r_s}|}.$$

Similarly, we can apply superposition and other sources:

$$V = \sum \frac{q_i}{4\pi\varepsilon r} + \frac{1}{4\pi\varepsilon}\int \frac{\rho_L}{r_o}dL' + \frac{1}{4\pi\varepsilon}\int \frac{\rho_S}{r_o}dS' + \frac{1}{4\pi\varepsilon}\int \frac{\rho_V}{r_o}dv'.$$

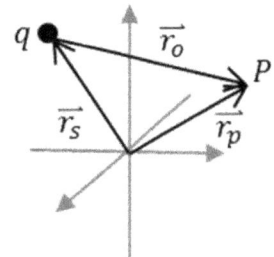

Finding electric potential using vector form.

The second method uses **Poisson's equation** derived from Guass's equation:

$$\nabla \cdot \vec{D} = \nabla \cdot \varepsilon\vec{E} = -\varepsilon[\nabla \cdot (\nabla V)] = \rho_V$$

$$\nabla^2 V = \frac{-\rho_V}{\varepsilon}.$$

Note: When no sources are present, Poisson's equation reduces to **Laplace's equation**: $\nabla^2 V = 0$.

Example Find the potential at a point (0,0,z) from a ring of surface charge with inner and outer radii a and b.

We'll use the adaptation on Coulomb's law for potential with:

$$r_o = \sqrt{r_s^2 + z^2}; \quad dS' = r_s \partial r_s \partial\theta;$$

$$V = \int_0^{2\pi} \int_a^b \frac{\rho_s r_s \partial r_s \partial\theta}{4\pi\varepsilon\sqrt{r_s^2 + z^2}}$$

$$= \frac{\rho_s}{2\varepsilon}\left[\sqrt{a^2 + z^2} - \sqrt{b^2 + z^2}\right] \text{ V}$$

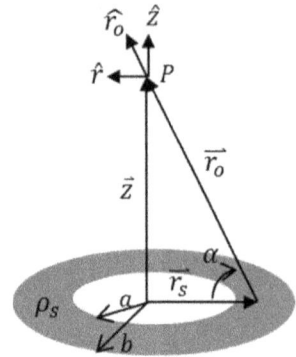

A charged ring.

Example Find an expression for the potential in a dielectric between two conductors where the inner conductor of radius a has a voltage of V_o and the outer conductor of radius b has a voltage of 0 V.

This is a good problem for Poisson (or Laplace). There is no charge in the dielectric, so we can use $\nabla^2 V = 0$.

From prior work, we know that there is only a cylindrical $\hat{r}$ dependence of the electric field (and by extension of potential) due to this type of geometry.

Dielectric between two conductors.

$$\nabla^2 V = \frac{1}{r}\frac{\partial}{\partial r}\left(r\frac{\partial V}{\partial r}\right) = 0$$

$$r\frac{\partial V}{\partial r} = A \rightarrow \frac{\partial V}{\partial r} = \frac{A}{r} \rightarrow V = A \ln r + B$$

$$V(a) = A \ln a + B = V_o;$$
$$V(b) = A \ln b + B = 0; \rightarrow B = -A \ln b;$$
$$A(\ln a - \ln b) = A \ln\frac{a}{b} = V_o \rightarrow A = \frac{V_o}{\ln a/b};$$

$$V = \frac{V_o}{\ln a/b}(\ln r - \ln b) = \frac{V_o \ln r/b}{\ln a/b} \text{ V}$$

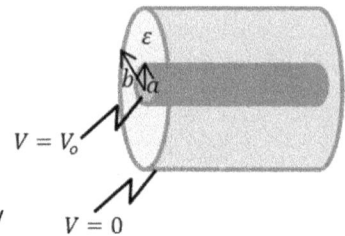

Electric Dipoles:

Electric dipoles are pairs of opposite but equal charge, +q and –q, separated by a small distance, d.

The potential due to a dipole can be found using superposition:

$$V = \sum \frac{q_i}{4\pi\varepsilon r} = \frac{q}{4\pi\varepsilon r_1} - \frac{q}{4\pi\varepsilon r_2} = \frac{q}{4\pi\varepsilon}\left[\frac{\Delta r}{r_1 r_2}\right]$$

Since d is small, if our observation point is far enough, we can say $r_1 \cong r_2 \cong r$ and $r_1 r_2 \cong r^2$. Also, from the geometry of the problem, we can find the difference between r_1 & r_2 which is $\Delta r = d\cos\theta$. This leaves us with:

$$V = \frac{qd\cos\theta}{4\pi\varepsilon r^2}.$$

We can define a **dipole moment**, $\vec{p}$, which indicates the power of the dipole in C·m. The dipole moment points from the negative to the positive charge and is given by:

$$\vec{p} = q\vec{d}.$$

By definition of the dot product, we can write that $\vec{p}\cdot\hat{r} = qd\cos\theta$ where $\hat{r}$ is the unit vector pointing in the direction to the observation point. Now we have:

$$V = \frac{\vec{p}\cdot\hat{r}}{4\pi\varepsilon r^2}.$$

Finally, we can find the electric field from the dipole by:

$$\vec{E} = -\nabla V = \frac{-\partial V}{\partial r}\hat{r} - \frac{1}{r}\frac{\partial V}{\partial\theta}\hat{\theta} - \frac{1}{r\sin\theta}\frac{\partial V}{\partial\varphi}\hat{\varphi}$$

$$\vec{E}_{dipole} = \frac{qd}{4\pi\varepsilon r^3}\left[2\cos\theta\,\hat{r} + \sin\theta\,\hat{\theta}\right].$$

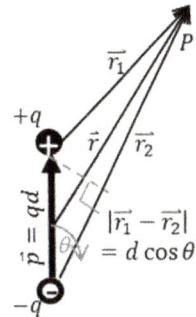

Geometry of a dipole: two charges, +q and –q, located a distance, d, from each other and observed at a point, P.

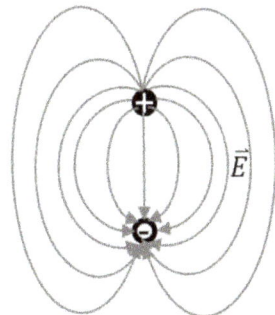

Electric field intensity of a dipole.

Chapter 4: Magneto-Statics

Current, Current Density, & Conductance:

The two main sources of the static magnetic field as stated in Maxwell's equations are **current** and **current density**.

Current is a scalar quantity that is a measure of the amount of charge flowing in a given time. The units are C/s or Amps (A). The current density is a vector quantity that points in the direction of current flow with units of A/m^2. The current density is the amount of charge that passes through a cross-sectional area in a given amount time.

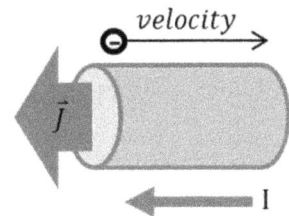

Current is the flow of charge. Current density is the flow of charge through a cross-sectional area.

Current and current density are related by the following general equation:

$$I = \int \vec{J} \cdot d\vec{S}.$$

Conductance, σ, is a property that is especially important when dealing with current and current density. It is a measure of how well charge can flow through a material and is given units of Siemens per meter (S/m) or Mho's per meter (1/Ω·m). In a perfect dielectric, $\sigma = 0$, and in a perfect conductor, $\sigma = \infty$.

Conductance is used in **Ohm's law** to relate the current density and electric field as follows:

$$\vec{J} = \sigma \vec{E}.$$

Using the above relation, we find the following is true of ideal materials:

Perfect Dielectric: $\sigma = 0;\quad \vec{J} = 0$
Perfect Conductor: $\sigma = \infty;\quad \vec{E} = 0$

Find the magnetic field due to a hollow cylinder with inner radius a and outer radius b carrying a current density of $\vec{J} = J\hat{z}$.

We'll use Ampère's law, $\oint \vec{H} \cdot d\vec{L} = I$.

$$d\vec{L} = r\partial\theta \; \hat{\theta}$$

$$\int_0^{2\pi} H_\theta \, r\partial\theta = 2\pi r H_\theta;$$

Now we need to find I:

$$I = \int \vec{J} \cdot d\vec{S}$$

$$d\vec{S} = r\partial r\partial\theta \; \hat{z}$$

Between the two walls of the cylinder:

$$\int_0^{2\pi} \int_a^r J \, r\partial r\partial\theta = \pi(b^2 - r^2)J$$

Outside the cylinder:

$$\int_0^{2\pi} \int_a^b J \, r\partial r\partial\theta = \pi(b^2 - a^2)J$$

And, of course, there is no current where $r < a$.

Putting it together we have:

$$2\pi r H_\theta = \pi(b^2 - r^2)J$$

$$\vec{H} = \frac{(b^2 - r^2)J}{2r} \; \hat{\theta} \quad \text{A/m} \;\; a < r < b$$

$$\vec{H} = \frac{(b^2 - a^2)J}{2r} \; \hat{\theta} \quad \text{A/m} \;\; r > a$$

$$\vec{H} = 0 \quad \text{A/m} \;\; r < a \;\; \text{(no current enclosed)}$$

Example

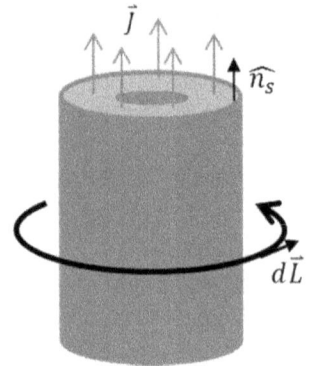

Current flowing through a hollow cylinder.

Permeability & Susceptibility:

Permeability, μ, is an expression of the effects a material has on the magnetic field which propagates through it. Permeability, like permittivity, can cause refraction, retardation, and attenuation. The materials that we will consider in this text will be described by a constant permittivity.

And, just as with permittivity, it is convenient to describe the **permeability of free space**, μ_0, by which all other materials are compared. The value of this quantity is:

$$\mu_0 = 4\pi \times 10^{-7} \text{ H/m.}$$

For a vacuum, air, and most conductors and dielectrics, the free space permeability adequately describes the material. However, in the presence of a magnetic field, **magnetic dipoles** can be induced. A magnetic dipole is an arrangement such as a bar magnet or current loop that causes a magnetic field which looks very similar to the electric field of an electric dipole. We can talk about a **magnetization field**, $\vec{M}$, that arises from these induced dipoles such that :

$$\vec{M} = \mu_0 \chi_m \vec{H}$$

where χ_m is the **magnetic susceptibility**. The magnetic flux density, $\vec{B}$, including the magnetization field becomes:

$$\vec{B} = \mu_0 \vec{H} + \mu_0 \chi_m \vec{H} = \mu_o (1 + \chi_m)\vec{H} \equiv \mu \vec{H}.$$

By fulfilling the constituent relation $\vec{B} = \mu \vec{H}$, we get that $\mu = \mu_0(1 + \chi_m)$. We can define a new term, the **relative permeability**, $\mu_r = (1 + \chi_m)$ which gives:

$$\mu = \mu_o(1 + \chi_m) = \mu_0 \mu_r.$$

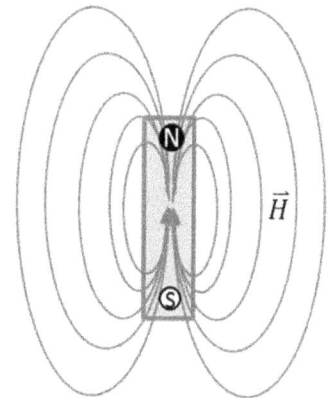

Magnetic field intensity from a bar magnet.

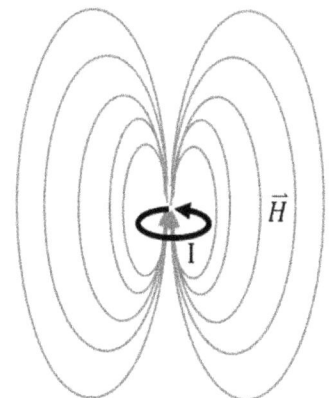

Magnetic field intensity from a current loop.

Biot-Savart's Law:

First of all, this is pronounced Bee-oh-sah-Var (capital letters showing stressed syllables). Unless you have studied French, there's no way to know that.

With that out of the way, similar to Coulomb's law, the **Biot-Savart law** gives us a way to find the magnetic fields when the sources are known. The equations themselves are fairly straightforward, but, as usual, it's the implementation that can be difficult. The Biot-Savart law states:

$$\vec{H} = \frac{1}{4\pi} \int \frac{I\, d\vec{L'} \times \hat{r}_o}{r_o{}^2} = \frac{1}{4\pi} \int \frac{\vec{J} \times \hat{r}_o}{r_o{}^2} \, dv'.$$

Just as with Coulomb's law, the integration is over the _source_. Unlike the line integrals of electric potential which yield a scalar value, the vector valued magnetic field does depend on the path the current is actually taking.

Find the magnetic field due to a current carrying infinite wire.

Example

This time, $d\vec{L} = \partial r_s\, \hat{z}$;

$$r_o{}^2 = r_s^2 + r^2;$$
$$\hat{r}_o = \sin\alpha\, \hat{r} - \cos\alpha\, \hat{z} = \frac{r}{\sqrt{r_s^2 + r^2}}\, \hat{r} - \frac{r_s}{\sqrt{r_s^2 + r^2}}\, \hat{z};$$

As $\hat{z} \times \hat{z} = 0$, we're left again with only $\hat{r}$ terms. We also know that $\hat{z} \times \hat{r} = \hat{\theta}$:

$$\vec{H} = \frac{I}{4\pi} \int_{-\infty}^{\infty} \frac{r\, \partial r_s}{\sqrt{r_s^2 + r^2}^3}\, \hat{\theta} = \frac{I\hat{\theta}}{4\pi r} \left(\frac{r_s}{\sqrt{r_s^2 + r^2}} \right) \Big|_{-\infty}^{\infty}$$

$$= \frac{I}{4\pi r} \left(\frac{\infty}{\infty} - \frac{-\infty}{\infty} \right) \hat{\theta} = \frac{I}{2\pi r}\, \hat{\theta} \ \text{A/m}$$

which is what we got using Ampère's law.

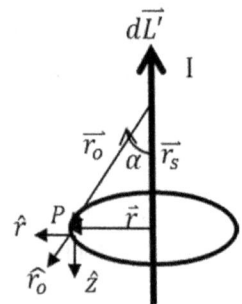

Geometry for evaluating magnetic field from a current carrying wire using Biot-Savart.

Example

Find the magnetic field due to a current loop with radius a.

We'll begin by identifying the main components for Biot-Savart in cylindrical coordinates:

$$\vec{dL'} = a\partial\theta \,\hat{\theta}$$
$$\vec{r} = r\cos\beta \,\hat{r} + r\sin\beta \,\hat{z};$$
$$\vec{r_o} = (r\cos\beta - a)\hat{r} + r\sin\beta \,\hat{z};$$

Using the law of cosines:
$$|\vec{r_o}|^2 = r^2 + a^2 - ra\cos\beta;$$

We also note that $\hat{\theta} \times \hat{r} = -\hat{z}$ and $\hat{\theta} \times \hat{z} = \hat{r}$. Putting it all together:

$$\vec{H} = \frac{I}{4\pi}\int_0^{2\pi} \frac{(a - r\cos\beta)\hat{z} + r\sin\beta\,\hat{r}}{\sqrt{r^2 + a^2 - ra\cos\beta}^3} \cdot a\,\partial\theta$$

$$= \frac{I}{2} \frac{a(a - r\cos\beta)\hat{z} + ar\sin\beta\,\hat{r}}{\sqrt{r^2 + a^2 - ra\cos\beta}^3} \text{ A/m}$$

Note that when $\beta = \pi$, we are on the z-axis and $r = z$:

$$\vec{H}(0,0,z) = \frac{I}{2}\frac{a^2\hat{z} + az\,\hat{r}}{\sqrt{z^2 + a^2}^3} \text{ A/m}$$

When $\beta = 0$, we are on the r-axis:

$$\vec{H}(r,0,0) = \frac{I}{2}\frac{a(a - r)\hat{z}}{\sqrt{r^2 + a^2 - ra}^3} \text{ A/m}$$

The latter result implies that when we are inside the loop, $\vec{H}$ points in the $+\hat{z}$ direction, whereas outside the loop, $\vec{H}$ points in the $-\hat{z}$ direction. This is similar to the result we obtained in spherical coordinates with Ampère's law which implies that $\vec{H}$ circles around the current loop.

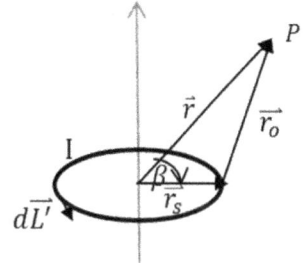

Geometry for evaluating magnetic field from a current carrying loop using Biot-Savart.

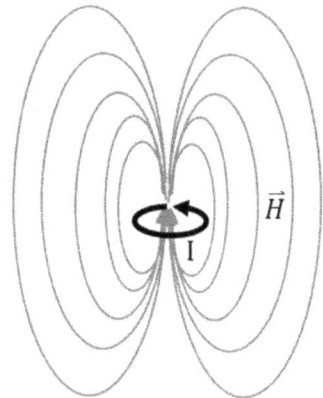

Resulting field.

Find magnetic field of a solenoid.

Example

A **solenoid** is a current carrying wire arranged in loops.

<u>Method 1:</u> The first way we'll do this is by recalling from the previous example that one current loop gives:

$$\vec{H}(0,0,z) = \frac{Ia^2}{2\sqrt{z^2+a^2}^3}\hat{z}$$

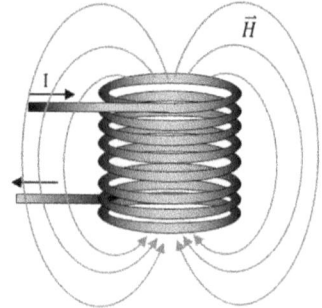

Solenoid with Field.

To get the contribution of N coils with a total of NI current over the length, we sum, or integrate, over the length, l:

$$NI = \frac{NI}{l}\int_0^l \partial z \rightarrow$$

$$\vec{H} = \frac{NI}{l}\int_0^l \frac{Ia^2\partial z}{2\sqrt{z^2+a^2}^3}\hat{z} = \frac{NI}{2l}\frac{l}{\sqrt{l^2+a^2}}\hat{z} \rightarrow \frac{NI}{2l}\hat{z}\ (l>>a).$$

<u>Method 2:</u> The second way we'll do it is to consider a cross-section of the solenoid where we draw a rectangular loop with diminishingly small dimensions in the $\pm\hat{r}$ directions:

$$\oint \vec{H}\cdot d\vec{L} = I \rightarrow \int_0^l H\partial z + \int_0^l(-H)(-\partial z) = NI$$

$$2lH = NI \rightarrow \vec{H} = \frac{NI}{2l}\hat{z}$$

Now we sum, or integrate, the total contribution over the circumference of the solenoid:

$$\frac{NI}{2\pi r}\int_0^{2\pi} r\partial\theta = NI \rightarrow \vec{H} = \int_0^{2\pi}\frac{NI}{2\pi lr}r\partial\theta\hat{z} = \frac{NI}{2l}\hat{z}$$

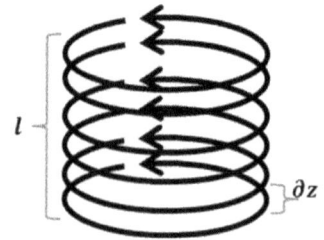

Geometry for Method 1.

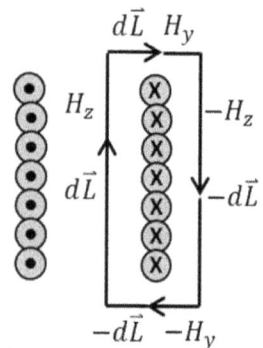

Geometry for Method 2.

I must point out that other sources claim another factor of 2. I've found either no derivation of their answer, an assumption that the magnetic field is zero outside the solenoid (not true!), or a math error in the derivation. Nonetheless, the widely accepted answer is $\vec{H} = {NI}/{l}\hat{z}$.

Magnetic Potential:

The **magnetic potential** is the magnetic counterpart to the electrical potential but in vector form. It is a conceptual tool most often used to simplify problems that are more difficult to solve with Ampère's law or Biot-Savart.

By definition, the magnetic potential, $\vec{A}$, has units of Webers per meter (Wb/m) and is related to the magnetic field density by:

$$\vec{B} = \nabla \times \vec{A}.$$

To comply with Maxwell's laws and to serve convenience, the magnetic potential must also obey:

$$\nabla \cdot \left(\nabla \times \vec{A} \right) = 0; \quad \nabla \cdot \vec{A} = 0.$$

To find a method to calculate magnetic potential, we start with Ampère's law written as:

$$\nabla \times \mu \vec{H} = \nabla \times \vec{B} = \nabla \times \left(\nabla \times \vec{A} \right) = \mu \vec{J}.$$

Using the vector relation $\nabla \times \left(\nabla \times \vec{A} \right) = \nabla(\nabla \cdot \vec{A}) - \nabla^2 \vec{A}$ and the constraint $\nabla \cdot \vec{A} = 0$, we get **Poisson's vector equation**:

$$\nabla^2 \vec{A} = -\mu \vec{J}.$$

Solving the above and converting to integral form, the magnetic potential is given by:

$$\vec{A} = \frac{\mu}{4\pi} \int \frac{I \, d\vec{L'}}{r_o} = \frac{\mu}{4\pi} \int \frac{\vec{J}}{r_o} dv'.$$

Example

Find the magnetic field inside a coaxial cable carrying a current, I.

Between the inner and outer conductors, there should be no current flow, therefore $\vec{J} = 0$ in the dielectric. However, because current flows in the $+\hat{z}$ direction, we assume $\vec{A} = A_z\hat{z}$. We are also going to assume that A_z only varies with r due to the symmetry of the problem. From Laplace:

$$\nabla^2 A_z = \frac{1}{r}\frac{\partial}{\partial r}\left(r\frac{\partial A_z}{\partial r}\right) = 0$$

$$r\frac{\partial A_z}{\partial r} = C \rightarrow A_z = C\ln r + D;$$

If we finally assume the magnetic potential on the outer surface at radius b is 0, then we have:

$$C\ln b + D = 0 \rightarrow D = -C\ln b \rightarrow \vec{A} = C\ln\frac{r}{b}\,\hat{z}$$

$$\mu\vec{H} = \nabla \times \vec{A} = \frac{1}{r}\frac{\partial A_z}{\partial\theta}\,\hat{r} - \frac{\partial A_z}{\partial r}\,\hat{\theta} = -\frac{C}{br}\,\hat{\theta}$$

Lastly, we use Ampère's law to find C:

$$\oint \vec{H}\cdot d\vec{L} = \int_0^{2\pi}\frac{-C}{\mu br}r\,d\theta = -\frac{2\pi C}{\mu b} = I \rightarrow C = \frac{-I\mu b}{2\pi}$$

$$\vec{H} = -\frac{C}{\mu br}\,\hat{\theta} = \frac{I}{2\pi r}\,\hat{\theta}\ \text{A/m}$$

Personally, I hate working with $\vec{A}$. Let's do the same problem using Ampère's law without all of the hand-waving:

$$\oint \vec{H}\cdot d\vec{L} = \int_0^{2\pi}H_\theta r\,d\theta = 2\pi r H_\theta = I$$

$$\vec{H} = \frac{I}{2\pi r}\,\hat{\theta}$$

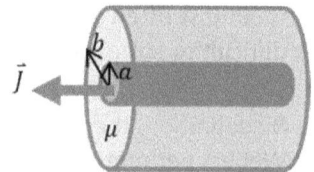

Coaxial cable carrying a current on the inner conductor.

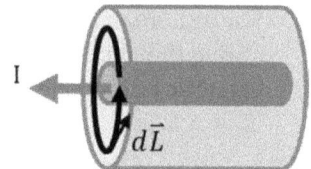

Coaxial cable with Ampère's law.

Magnetic Dipoles:

The **magnetic dipole** consists of two magnetic monopoles, north and south, of equal magnitude and opposite polarity separated by a small distance. We can define a **magnetic dipole moment**, $\vec{m}$, pointing from south to north which indicates the power of the dipole in A·m². For a current loop with radius a, the magnetic dipole moment is by:

$$\vec{m} = I\pi a^2 \hat{z}.$$

Expressing $\vec{m}$ in spherical coordinates, we can write:

$$\vec{m} = m(\cos\theta\,\hat{r} + \sin\theta\,\hat{\theta}).$$

The magnetic potential of a magnetic dipole is defined as:

$$\vec{A} = \frac{\mu}{4\pi r^2}(\vec{m} \times \hat{r}).$$

Performing the cross product yields:

$$\vec{A} = \frac{\mu m \sin\theta}{4\pi r^2}\,\hat{\varphi}$$

Finally, we use the relation $\vec{H}/\mu = \nabla \times \vec{A}$ which in spherical coordinates yields:

$$\vec{H}_{dipole} = \frac{m}{4\pi r^3}\left[2\cos\theta\,\hat{r} + \sin\theta\,\hat{\theta}\right].$$

Note the similarity to the electric dipole:

$$\vec{E}_{dipole} = \frac{qd}{4\pi\varepsilon r^3}\left[2\cos\theta\,\hat{r} + \sin\theta\,\hat{\theta}\right].$$

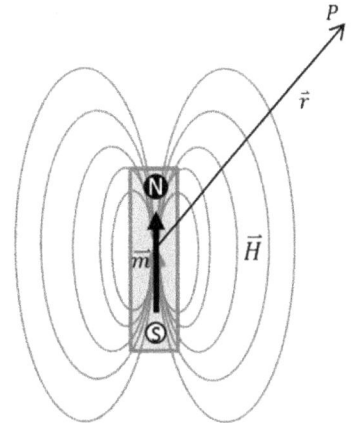

Magnetic field lines from a magnetic dipole with moment, $\vec{m}$, observed at a point, P.

Chapter 5: Boundary Conditions

Normal Boundary Conditions:

When two dielectrics meet in space, we want to know what happens to the electro-magnetic field across the boundary. Consider the geometry in the figure at the right where two materials meet. There may exist a surface charge density, ρ_s, on that boundary.

We can break the electric and magnetic fields into normal and tangential components as follows:

$$\vec{E} = E_n \hat{n} + E_t \hat{t}; \quad \vec{H} = H_n \hat{n} + H_t \hat{t}$$

where $\hat{n}$ and $\hat{t}$ are unit vectors normal and tangential to the boundary surface, respectively.

Geometry to find the normal components of the electro-magnetic field at the boundary of two dielectrics.

We will next use a cylinder with height, Δh, and radius, a, to describe a closed surface that straddles the boundary. Using Gauss' law for electro-statics as $\Delta h \to 0$, we have:

$$\oint \vec{D} \cdot d\vec{S} = \int \vec{D_1} \cdot d\vec{S_1} + \int \vec{D_2} \cdot d\vec{S_2} = Q = \rho_s \, \pi a^2.$$

Using $d\vec{S_1} = -d\vec{S_2} = r \, d\theta \, \hat{n}$, where $\hat{n} \equiv \widehat{n_1} = -\widehat{n_2}$:

$$D_{n1} \, \pi a^2 - D_{n2} \, \pi a^2 = \rho_s \, \pi a^2$$

$$D_{n1} - D_{n2} = \rho_s; \quad \varepsilon_1 E_{n1} - \varepsilon_2 E_{n2} = \rho_s.$$

Similarly, we can use Gauss' law for magneto-statics to obtain:

$$\oint \vec{B} \cdot d\vec{S} = \int \vec{B_1} \cdot d\vec{S_1} + \int \vec{B_2} \cdot d\vec{S_2} = 0.$$

$$B_{n1} = B_{n2}; \quad \mu_1 H_{n1} = \mu_2 H_{n2}.$$

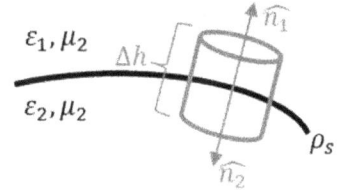

Tangential Boundary Conditions:

Similar to how we found what happens to the normal components of the electro-magnetic field at a boundary, we can also find the tangential components.

We can draw a rectangular loop of width, ΔL, and height, Δh, that straddles the boundary between two dielectrics. Using Kirchhoff's law and letting $\Delta h \to 0$, we find:

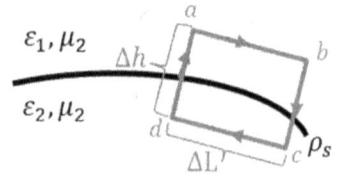

$$\oint \vec{E} \cdot d\vec{L} = \int_a^b \overrightarrow{E_1} \cdot d\overrightarrow{L_1} + \int_c^d \overrightarrow{E_2} \cdot d\overrightarrow{L_2} = 0.$$

From the sketch, we know that $d\overrightarrow{L_1} = -d\overrightarrow{L_2} = \Delta L \hat{t}$. Now we have the relation:

Geometry to find the tangential components of the electro-magnetic field at the boundary of two dielectrics.

$$E_{t1} \Delta L - E_{t2} \Delta L = 0$$

$$E_{t1} = E_{t2}; \quad \varepsilon_1 D_{t1} = \varepsilon_2 D_{t2}.$$

Performing a similar operation using Ampère's law yields:

$$\oint \vec{H} \cdot d\vec{L} = \int_a^b \overrightarrow{E_1} \cdot d\overrightarrow{L_1} + \int_c^d \overrightarrow{E_2} \cdot d\overrightarrow{L_2} = \mathrm{I}.$$

If we picture a current flowing through the loop, we get:

$$\mathrm{I} = \int \vec{J} \cdot d\vec{S} = J\,(\Delta h \Delta L) \to 0$$

which you might expect since we are not dealing with conductors and therefore not really dealing with mobile electrons. With that in mind, we get the final relationships:

$$H_{t1} = H_{t2}; \quad \frac{B_{t1}}{\mu_1} = \frac{B_{t2}}{\mu_2}.$$

Example

Find the electric and magnetic fields after a dielectric-dielectric boundary where $\varepsilon_2 = 2\varepsilon_1$ and $\mu_2 = \mu_1 = \mu_0$ if the fields in the first dielectric are given by:

$$\overrightarrow{E_1} = E_1 \sin\theta \, \hat{x} - E_1 \cos\theta \, \hat{y}$$

$$\overrightarrow{H_1} = H_1 \sin\theta \, \hat{x} - H_1 \cos\theta \, \hat{y}$$

Using boundary conditions:

$$E_{t2} = E_{t1} = E_1 \sin\theta;$$

$$E_{n2} = \frac{\varepsilon_1}{\varepsilon_2} E_{n1} = -\frac{1}{2} E_1 \cos\theta;$$

$$H_{t2} = H_{t1} = H_1 \sin\theta;$$

$$H_{n2} = \frac{\mu_1}{\mu_2} H_{n1} = -H_1 \cos\theta;$$

Putting it together, we have:

$$\overrightarrow{E_2} = E_1 \sin\theta \, \hat{x} - \frac{1}{2} E_1 \cos\theta \, \hat{y} \quad \text{V/m}$$

$$\overrightarrow{H_2} = H_1 \sin\theta \, \hat{x} - H_1 \cos\theta \, \hat{y} \quad \text{A/m}$$

The magnetic field passes as though there is no boundary, and the electric field shrinks in magnitude in the normal direction.

The resulting angle for the electric field in the second medium is given by the relation:

$$\tan\theta_2 = \frac{E_1 \sin\theta}{(\varepsilon_1/\varepsilon_2) E_1 \cos\theta} = \frac{\varepsilon_2}{\varepsilon_1} \tan\theta = 2 \tan\theta$$

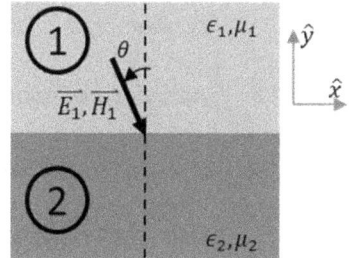

Geometry for the electro-magnetic field at the boundary of two dielectrics.

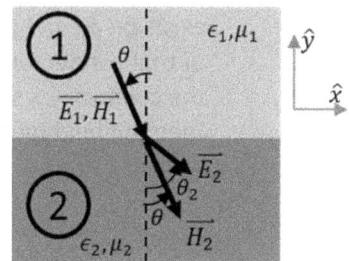

The resulting fields after the boundary of two dielectrics.

Example

Determine the electric field just outside a planar conductor.

Recall from Chapter 4 that $\vec{J} = \sigma \vec{E}$. In a perfect conductor, $\sigma = \infty$, so $\overrightarrow{E_2} = 0$.

Applying boundary conditions, we have:

$$E_{t1} = E_{t2} = 0; \quad \varepsilon_1 E_{n1} - \varepsilon_1 E_{n1} = \varepsilon_1 E_{n1} = \rho_s;$$

Using cartesian coordinates with $\hat{t}$ on the x-y plane and $\hat{n} = \hat{z}$:

$$\overrightarrow{E_1} = \frac{\rho_s}{\varepsilon_1} \hat{z}$$

We note first that this is exactly twice as large as the result we obtained for the infinite thin sheet. In that case, our electric field could emit from either surface. Here, the electric field can only propagate away from the conductor on one side of the boundary.

The second thing to note is that if a surface charge exists on a conductor, the electric field just outside of that conductor is perpendicular everywhere to the conductor's surface. If no surface charge exists, then the field right at the conductor surface is 0 V/m.

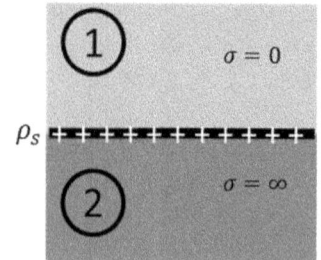

Geometry for the electric field at the boundary of a perfect conductor.

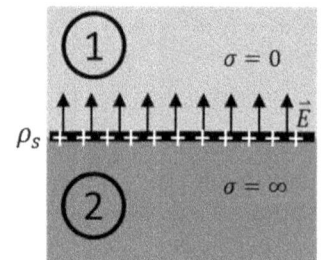

The resulting field after the conductor boundary.

Image Theory:

So far, we've considered the electric field found from multiple sources and what occurs to the electric field at dielectric boundaries and at a conductor boundary with no other sources present. If we now place a source near a dielectric-conductor boundary, then the analysis changes.

If a source exists near a conductor, the free electrons of the conductor will respond by setting up a corresponding surface charge. From the previous example, we learned that the electric field next to a conductor with a surface charge is perpendicular to the surface. Since $\vec{E} = E_z \hat{z}$ and $d\vec{L} = dx\hat{x} + dy\hat{y}$, we know that $V = -\int \vec{E} \cdot d\vec{L} = 0$ everywhere on the conductor surface. In other words, the conductor surface is an **equipotential surface** meaning $V_{ab} = 0$ for all points.

Various sources next to a perfect conducting plane.

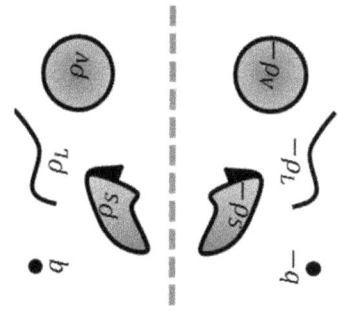

We can accommodate such a restriction if we treat the conducting plane as an inversion mirror. In **image theory**, the conduction plane is replaced by a mirror of sources of equal but opposite charge as those that exist outside the conduction plane.

The electric field is now found by summing the contribution from all of the charges, both real and virtual, using Coulomb's law and the law of linear superposition.

To check if $V_{ab} = 0$, we recall the equation for potential as:

$$V_{ab} = V_b - V_a = -\int_a^b \vec{E} \cdot d\vec{L}.$$

Because the electric field from the image source is equal and opposite the field from the original source, and because both sources are equidistant from the image plane, we get that indeed $V_{ab} = 0$ everywhere on the plane where the conductor once was.

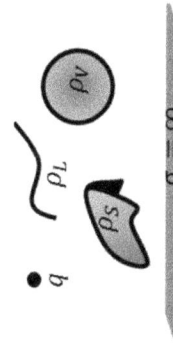

Using image theory replaces the conduction plane with equal and opposite images of the original charges.

41

Example

Find the electric field from a charge of $+q$ located a distance $d/2$ from a conducting plane.

We replace the conducting plane with the image of the source and find that we have the geometry of a dipole. We already know the electric field from our prior evaluations:

$$\vec{E}_{dipole} = \frac{qd}{4\pi\varepsilon r^3}\left[2\cos\theta\,\hat{r} + \sin\theta\,\hat{\theta}\right] \text{ V/m}$$

Note that the electric field lines are perpendicular to the image plane where the conductor used to be.

To check, we take the case where in spherical coordinates $\theta = \pi$ which puts us on the x-y plane. There the electric field has only a $\hat{\theta}$ term which can be converted to Cartesian coordinates:

$$\hat{\theta} = \cos\theta\cos\varphi\,\hat{x} + \cos\theta\sin\varphi\,\hat{y} - \sin\theta\,\hat{z}$$

$$= 0\cdot\cos\varphi\,\hat{x} + 0\cdot\sin\varphi\,\hat{y} + \hat{z} = \hat{z}$$

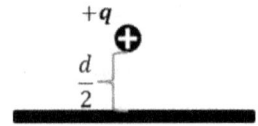

A point charge next to a perfect conducting plane.

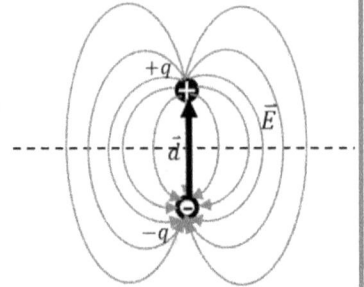

Image theory transformation of the problem. A point charge close to a conducting surface acts as a dipole.

Chapter 6: Circuit Measurements

Resistance, Conductance, & Capacitance:

We've already discussed two entities that can easily be measured in an electric circuit: current and voltage. It should be no surprise that we can also talk about the other major players in electric circuits: resistance, conductance, capacitance, and inductance. The first three attributes deal with the electric field, so we'll consider them first.

Resistance, R, is a measurement of how difficult it is to mobilize a free charge in a volume. It is measured in ohms (Ω) or Volts/Ampère (V/A) and is given by:

$$R = \frac{V}{I} = \frac{-\int \vec{E} \cdot d\vec{L}}{\int \sigma \vec{E} \cdot d\vec{S}}.$$

Conductance, G, is simply the inverse of resistance measured in Mho's ($1/\Omega$) or Siemens (S). It measures the ease of moving a charged particle in a material.

$$G = \frac{1}{R} = \frac{I}{V}.$$

Capacitance, C, is a phenomenon that occurs when a dielectric is placed between two conductors. When a potential is applied to the system, the conductors respond by collecting equal and opposite charge along their surfaces as shown to the right. In turn, the molecules and atoms of the dielectric polarize. The ratio of the induced surface charge of the conductor with the potential of the electric field across the dielectric is the capacitance in Farads (F) or Coulombs/Volt (C/V):

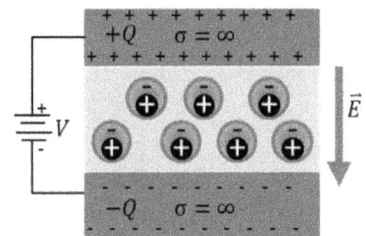

$$C = \frac{Q}{V} = \frac{\int \varepsilon \vec{E} \cdot d\vec{S}}{-\int \vec{E} \cdot d\vec{L}}.$$

A dielectric sandwiched between two conductors with a potential applied.

Because $\vec{E}$ appears in both numerator and denominator in R, G, and C calculations, these attributes are dependent on the system geometry and material properties and never on $\vec{E}$.

Example

Find the capacitance of a dielectric between two parallel plate conductors.

We know from prior examples that $\vec{E} = -E\hat{z}$.

$$Q = \int \varepsilon E \, dx \, dy = \varepsilon E \, A \quad (A = \text{plate area})$$

$$V = -\int E \, dz = E \, d \quad (d = \text{dielectric thickness})$$

$$C = \frac{Q}{V} = \frac{\varepsilon E A}{E d} = \frac{\varepsilon A}{d} \text{ which is independent of } \vec{E}.$$

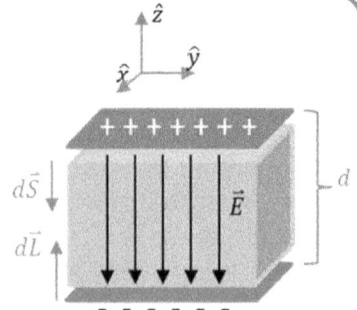

A parallel plate capacitor: Note $d\vec{S}$ points into the system and $d\vec{L}$ points to $+\hat{z}$.

Example

Find the capacitance and resistance of a coaxial cable with a non-perfect dielectric.

We know that in the dielectric shown, $\vec{E} = \frac{\rho_L}{2\pi\varepsilon r}\hat{r}$.

$$Q = \int_0^l \int_0^{2\pi} \varepsilon \frac{-\rho_L}{2\pi\varepsilon r} r \, \partial\theta \, \partial z = -l\rho_L \quad (l = \text{length})$$

Note that $d\vec{S}$ points into the system in $-\hat{r}$.

$$V = -\int_a^b \frac{\rho_L}{2\pi\varepsilon r} \partial r = \frac{-\rho_l}{2\pi\varepsilon} \ln\frac{b}{a}$$

$$C = \frac{Q}{V} = \frac{2\pi l\varepsilon}{\ln(b/a)} \text{ F}$$

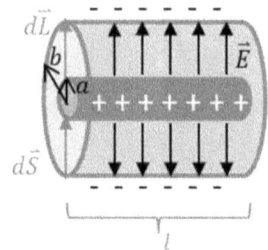

A coaxial cable: Note that $d\vec{S}$ points into the system and $d\vec{L}$ points in the $+\hat{r}$ direction.

For the resistance, we assume a non-zero σ and a current flow perpendicular to the inner conductor:

$$I = \int_0^l \int_0^{2\pi} \frac{-\sigma\rho_L}{2\pi\varepsilon r} r \, \partial\theta \, \partial z = \frac{-\sigma l\rho_L}{\varepsilon}$$

$$R = \frac{V}{I} = \frac{\ln^{b}/_a}{2\pi l\sigma} \, \Omega$$

Note that $RC = {}^{\varepsilon}/_{\sigma}$

Flux & Inductance:

As mentioned in the previous section, we can also determine formulations for inductance. To begin, we'll first define a parameter called **magnetic flux**, Φ, which is the amount of magnetic flux density, $\vec{B}$, transmitted through a given area. It has units of Webers (Wb) and is given by:

$$\Phi = \int \vec{B} \cdot d\vec{S} = \oint \vec{A} \cdot d\vec{L}.$$

A similar term called **magnetic flux linkage** or total magnetic flux, Λ, also measured in Wb, characterizes a system with N coils of current and is given simply by:

$$\Lambda = N\Phi = N \int \vec{B} \cdot d\vec{S} = N \oint \vec{A} \cdot d\vec{L}.$$

Finally, we can talk about **inductance**, L, which is the ratio of the total magnetic flux to the current in the system measured in Henries (H) or Webers per Ampere (Wb/A).

In the case of a single current source, the total magnetic flux through an area is due to that sole source. In such a case, we have **self-inductance** given by:

$$L = \frac{N}{I} \int \vec{B} \cdot d\vec{S}.$$

In the case where more than one current source exists, the sources act on themselves as well as with each other. To describe the interaction between sources, or the flux induced (thus inductance!) from one source to another, we have the **mutual inductance** given by:

$$L_{12} = \frac{N_2}{I_1} \int \vec{B_1} \cdot d\vec{S_2}.$$

where we integrate the magnetic flux density from source 1 over the area and number of current coils for source 2.

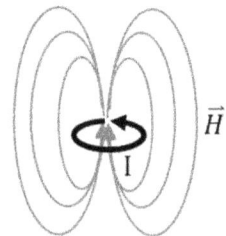

Self-inductance from a small current loop.

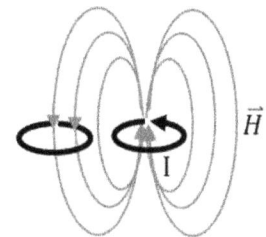

Mutual inductance imparted to a circle of wire from a small current loop.

Find the inductance inside a coaxial cable.

Example

We know from prior work that

$$\oint \vec{H} \cdot d\vec{L} = I \rightarrow \vec{H} = \frac{I}{2\pi r}\hat{\theta}$$

To find the flux, we look at a rectangular cross section perpendicular to the conductor:

$$d\vec{S} = \partial r\,\partial z\,\hat{\theta}$$

The flux is (with N=1):

$$\Phi = \int \vec{B} \cdot d\vec{S} = \int_0^l \int_a^b \frac{\mu I}{2\pi r}\,\partial r\,\partial z = \frac{\mu Il}{2\pi}\ln\frac{b}{a} \text{ Wb}$$

The inductance is

$$L = \frac{N\Phi}{I} = \frac{\mu l}{2\pi}\ln\frac{b}{a} \text{ H}$$

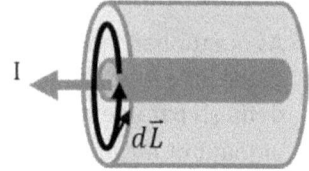

Magnetic field line in a coaxial cable.

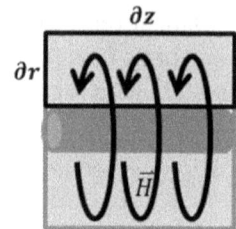

Flux lines in a coaxial cable.

Example

Find the inductance of a **toroid** (looped solenoid) with a square cross-sectional area wrapped by N coils of wire carrying a current I.

$$\oint \vec{H} \cdot d\vec{L} = I \rightarrow \int_0^{2\pi} -\vec{H} \cdot r\partial\theta\hat{\theta} = NI$$

$$\vec{H} = \frac{-NI}{2\pi r}\hat{\theta}$$

$$d\vec{S} = -\partial r\partial z\,\hat{\theta} \text{ (points with } \vec{H})$$

$$\Phi = \int \vec{B} \cdot d\vec{S} = \int_0^h \int_a^b \frac{\mu NI}{2\pi r}\,\partial r\,\partial z = \frac{\mu NIh}{2\pi}\ln\frac{b}{a}$$

$$L = \frac{N\Phi}{I} = \frac{\mu h N^2}{2\pi}\ln\frac{b}{a} \text{ H}$$

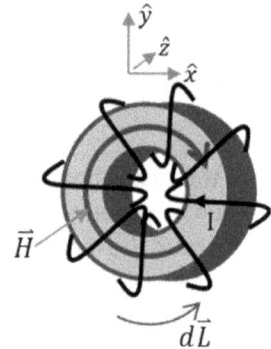

Magnetic field in a toroid.

Example

Find the mutual inductance due to arrangement shown.

The first thing to do is find the magnetic field inside the solenoid. For the sake of argument, we'll go with the widely accepted term:

$$\overrightarrow{B_1} = \frac{\mu N_1 I_1}{l} \hat{z}$$

The surface of coil 2 is:

$$d\vec{S} = r \, \partial\theta \, dr \, \hat{z}$$

The mutual inductance from coil 1 to coil 2 is:

$$L_{12} = \frac{N_2}{I_1} \int_0^{2a} \int_0^{2\pi} \frac{\mu N_1 I_1}{l} \hat{z} \cdot r \, \partial\theta \, dr \, \hat{z}$$

$$= \frac{4\pi \mu N_1 N_2 a^2}{l} \hat{z} \quad H$$

Going the other way:

$$\overrightarrow{B_2} = \frac{\mu N_2 I_2}{l} \hat{z}$$

$$d\vec{S} = r \, \partial\theta \, dr \, \hat{z}$$

$$L_{12} = \frac{N_1}{I_2} \int_0^{a} \int_0^{2\pi} \frac{\mu N_2 I_2}{l} \hat{z} \cdot r \, \partial\theta \, dr \, \hat{z}$$

$$= \frac{\pi \mu N_1 N_2 a^2}{l} \hat{z} \quad H$$

As you can see, having a radius differential of 2 between the coils leads to a 4x change in mutual inductance.

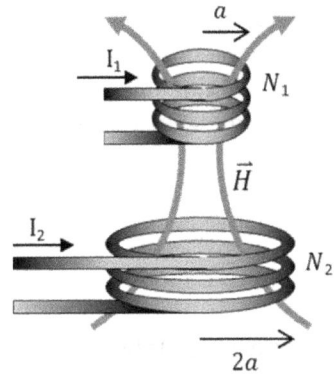

Mutual inductance problem.

Power & Energy:

In an electric circuit, we often are concerned with how much power, P, is dissipated in an element. That quantity is governed by the relation $P = V \cdot I$. Similarly in electro-magnetics, one can derive the **power**, P, dissipated in a volume which is given by **Joule's Law**:

$$P = \int \vec{E} \cdot \vec{J}\, dv = \int \sigma |\vec{E}|^2\, dv.$$

Power is measured in Watts (W).

In electro-statics, we can also talk about potential energy, particularly when dealing with the capacitor where energy is stored in the potential across the dielectric. The **electro-static potential energy**, W_e, measured in Joules (J), is given by:

$$W_e = \int \varepsilon |\vec{E}|^2\, dv.$$

Finally, there is a potential energy associated with the magnetic field, as well. In the case of an inductor, the current in the inductor gives rise to energy storage. The **magneto-static potential energy**, W_m, again measured in Joules (J), is given by:

$$W_m = \int \mu |\vec{H}|^2\, dv.$$

Example ➤ Find the electric and magnetic energy as well as dissipated power in a coaxial cable.

For electric energy:

$$I = \int_0^{2\pi} \int_0^l \vec{J} \cdot r\partial z\partial\theta \ \hat{r} = J2\pi rl \ \hat{r}$$

$$\vec{J} = \frac{I}{2\pi rl}\hat{r} = \sigma\vec{E} \rightarrow \vec{E} = \frac{I}{2\pi\sigma rl}\hat{r}$$

$$W_e = \int \varepsilon |\vec{E}|^2 dv$$

$$= \int_0^l \int_0^{2\pi} \int_a^b \frac{\varepsilon I^2}{4(\pi\sigma rl)^2} r\partial r\partial\theta\partial z = \frac{\varepsilon I^2}{2\pi\sigma^2 l}\ln\frac{b}{a} \quad \text{J}$$

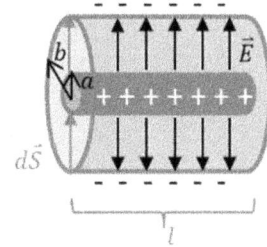

Coaxial cable with electric field shown. Here current goes from inner to outer conductor.

For magnetic energy:

$$\oint \vec{H} \cdot d\vec{L} = I \rightarrow \vec{H} = \frac{I}{2\pi r}\hat{\theta}$$

$$W_m = \frac{1}{2}\int \mu |\vec{H}|^2 \ dv$$

$$= \int_0^l \int_0^{2\pi} \int_a^b \frac{\mu I^2}{(2\pi r)^2} r\partial r\partial\theta\partial z = \frac{\mu l I^2}{2\pi}\ln\frac{b}{a} \quad \text{J}$$

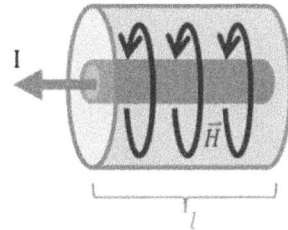

Coaxial cable with magnetic field shown.

Finally, for power dissipated:

$$P = \int \sigma |\vec{E}|^2 dv$$

$$= \int_0^l \int_0^{2\pi} \int_a^b \frac{I^2}{\sigma(2\pi rl)^2} r\partial r\partial\theta\partial z = \frac{I^2}{2\pi\sigma l}\ln\frac{b}{a} \quad \text{W}$$

Chapter 7: Moments, Forces, & Torque

Moments:

We have already covered two types of moments: the electric dipole moment and the magnetic dipole moment. Both of these moments measure the strength of the dipole.

For review, the **electric dipole moment**, $\vec{p}$, measured in C·m, is given by:

$$\vec{p} = q\vec{d}$$

where q is the magnitude of an individual charge in the dipole, and $\left|\vec{d}\right|$ is the separation between the charged pairs.

The electric field that arises from the electric dipole is:

$$\vec{E}_{dipole} = \frac{qd}{4\pi\varepsilon r^3}\left[2\cos\theta\,\hat{r} + \sin\theta\,\hat{\theta}\right].$$

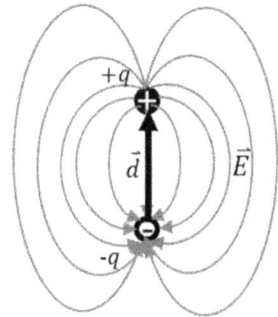

An electric dipole.

The magnetic dipole is characterized by a bar magnet or a small current loop. In fact, one can think of the bar magnet as having multiple small current loops within it. The **magnetic dipole moment**, $\vec{m}$, measured in A·m² is given by:

$$\vec{m} = I\pi a^2\hat{n}$$

where a is the radius of a loop with current I and $\hat{n}$ is the unit normal vector to the current loop.

The magnetic field that arises from the magnetic dipole is:

$$\vec{H}_{dipole} = \frac{m}{4\pi r^3}\left[2\cos\theta\,\hat{r} + \sin\theta\,\hat{\theta}\right].$$

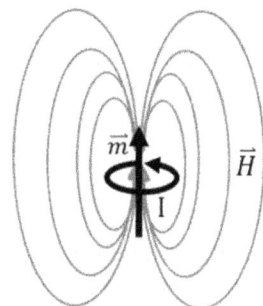

A magnetic dipole.

There are two other moments to consider when looking at the atomic level of a material. In classical physics, the electrons orbit a nucleus at some radius, r, and with some speed, $v = \pi r / T$ with T being the time to make one orbit. The current (charge divided by time) is then

$$I = (-e)v/\pi r.$$

The corresponding **orbital magnetic moment**, $\vec{m}_o$, of the electron in A·m^2 is then:

$$\vec{m}_o = I A \,\hat{n} = \frac{-evr}{2}\,\hat{n}.$$

A more correct formulation relates the orbital moment to the **angular momentum**, L, which is given by

$$L = m_e vr = 0, \hbar, 2\hbar, 3\hbar, \dots n\hbar$$

where m_e is the mass of the electron and $\hbar$ is Planck's constant divided by 2π. Substituting this into the above relation, we get:

$$\vec{m}_o = \frac{-en\hbar}{2m_e}\,\hat{n}.$$

A similar argument can be made for the **spin magnetic moment**, $\vec{m}_s$, also in A·m^2 which is given by:

$$\vec{m}_s = \frac{-e\hbar}{2m_e}\,\hat{n}.$$

It is interesting to note that in typical materials, the orientation of the individual atoms is random, thus cancelling the effects of the orbital magnetic moment on the macro-scale. As for spin magnetic moments, each atom with even numbered electrons has pairs with equal and opposite spin. Odd numbered electron effects also wash out on the macro-scale with the random orientation distributed in the material.

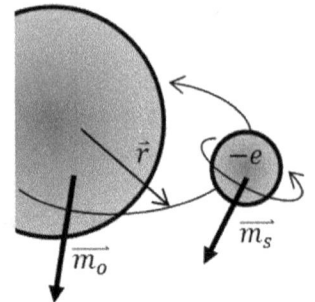

A spinning electron orbiting a nucleus gives rise to a spin magnetic moment and an orbital magnetic moment.

Coulomb Force:

The first part of **Coulomb's law** describes the electric field intensity due to sources of charge. The second part describes the electric force that arises due to that electric field.

According to the second part of Coulomb's law, the **electric force** or **Coulomb force**, $\vec{F_e}$, on a test charge, q', due to an already existent electric field, $\vec{E}$, is given by:

$$\vec{F_e} = q'\vec{E}$$

where force is measured in Newtons (N).

The electric force is related to the electro-static potential energy, W_e, by:

$$\vec{F_e} = -\nabla W_e.$$

Example

Find the force in a capacitor between the conducting plates.

As we derived before, the electric field at a dielectric-conductor boundary is:

$$\vec{E} = \frac{\rho_s}{\varepsilon}\hat{z}$$

$$W_e = \int_0^z \int_0^y \int_0^x \frac{\varepsilon \rho_s^2}{\varepsilon^2}\partial x \partial y \partial z = \frac{\rho_s^2}{\varepsilon}Az \quad \text{A=Area}$$

$$\vec{F_e} = -\nabla W_e = -\frac{\partial}{\partial z}\left(\frac{\rho_s^2}{\varepsilon}Az\right)\hat{z} = \frac{-A\rho_s^2}{\varepsilon}\hat{z} \quad \text{N}$$

We assumed the direction ($\hat{z}$) of the force based on the direction of the electric field.

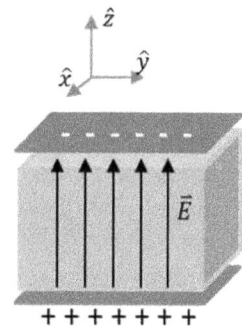

Parallel plate capacitor.

Lorentz Force:

Similar to placing a test charge particle in an electric field, we can also place the test charge in a magnetic field. If the particle is moving at a velocity, $\vec{v}$, the resulting **magnetic force**, $\overrightarrow{F_m}$, experienced by the particle in Newtons is given by:

$$\overrightarrow{F_m} = q'\vec{v} \times \vec{B}.$$

In the case of a current carrying wire in a magnetic field, the magnetic force is given by:

$$\overrightarrow{F_m} = I \int d\vec{L} \times \vec{B}.$$

The magnetic force is related to the magneto-static potential energy, W_m, by:

$$\overrightarrow{F_m} = -\nabla W_m.$$

With the electric and magnetic forces defined, we can finally define the **Lorentz force**, $\vec{F}$, which is simply the sum of the two forces:

$$\vec{F} = \overrightarrow{F_e} + \overrightarrow{F_m} = q'(\vec{E} + \vec{v} \times \vec{B}).$$

From the equations for electric and magnetic force, there are a few differences between the two which should be noted: For one, the electric force acts in the direction of the electric field, whereas the magnetic force acts perpendicular to the magnetic field. Also, an electric force can act on a stationary or moving charge; the magnetic force only affects a charge in motion.

Finally, and more subtly, the electric field can alter a particles trajectory and speed, but the magnetic field can only affect a particles trajectory as it does not impart energy to the system.

Find the Lorentz force in a coaxial cable.

Example

For the electric force, we have:

$$\vec{F_e} = -\nabla W_e = \frac{-\partial}{\partial r}\left(\frac{\varepsilon I^2}{2\pi\sigma^2 l}\ln r\right)\hat{r} = \frac{-\varepsilon I^2}{2\pi r\sigma^2 l}\hat{r}$$

We could also calculate this as follows:

$$I = \int_0^l \int_0^{2\pi} \frac{-\sigma\rho_L}{2\pi\varepsilon r}r\,\partial\theta\,\partial z = \frac{-\sigma l\rho_L}{\varepsilon} = \frac{-\sigma Q}{\varepsilon} = 2\pi r l\vec{J}$$

$$Q = -\frac{\varepsilon I}{\sigma}; \quad \vec{J} = \frac{I}{2\pi r l}\hat{r} = \sigma\vec{E} \rightarrow \vec{E} = \frac{I}{2\pi r l\sigma}\hat{r}$$

$$\vec{F_e} = q\vec{E} = \frac{-\varepsilon I^2}{2\pi r\sigma^2 l}\hat{r}$$

Coaxial cable with electric field shown. Here current goes from inner to outer conductor.

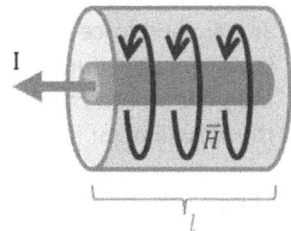

For the magnetic force:

$$\vec{F_m} = -\nabla W_m = \frac{-\partial}{\partial r}\frac{\mu l I^2}{2\pi}\ln r\,\hat{r} = \frac{-\mu l I^2}{2\pi r}\hat{r}$$

Or we could say:

$$I = \frac{Q}{l}\vec{v} \rightarrow Q\vec{v} = Il\hat{z}; \quad \vec{B} = \frac{\mu I}{2\pi r}\hat{\theta}$$

$$\vec{F_m} = Q\vec{v} \times \vec{B} = \frac{-\mu l I^2}{2\pi r}\hat{r}$$

Coaxial cable with magnetic field shown.

Total force is the sum of the two.

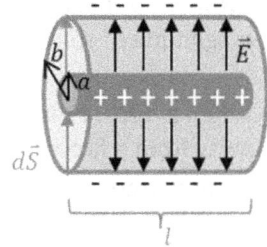

Find the force of wire 1 on wire 2.

Example

$$\vec{B_1} = \frac{\mu I_1}{2\pi r}\hat{\theta} = \frac{-\mu I_1}{2\pi d}\hat{x}; \quad Q_2\vec{v_2} = lI_2\hat{z};$$

$$\vec{F_{12}} = Q_2\vec{v_2} \times \vec{B_1} = \frac{-\mu l I_1 I_2}{2\pi d}\hat{y}$$

The wires carrying current in the same direction repulse each other.

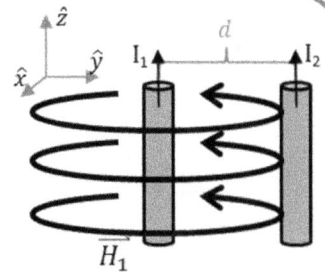

Forces between wires.

Torque:

The final topic we'll cover in electro-magneto-statics is **torque**. Torque is an expression of the direction and magnitude of spin about an axis when a force is applied.

When a force, $\vec{F}$, is applied to an object at a distance, $\vec{d}$, away from a fixed axis, the object will experience a torque, $\vec{T}$, measured in N·m, of:

$$\vec{T} = \vec{d} \times \vec{F}.$$

The vector, $\vec{d}$, called the **moment arm**, points from the axis to the point of applied force. The greater the moment arm and/or the force applied, the greater the torque.

The torque vector points perpendicular the moment arm and applied force and obeys the right hand rule. If you curl the fingers of your right hand from the direction of the moment arm to the direction of the applied force, your thumb will point in the direction of the torque vector.

In magneto-statics, the magnetic dipole can experience torque in the presence of a magnetic field. The magnetic dipole moment, $\vec{m}$, becomes the moment arm, and the magnetic flux density, $\vec{B}$, is the applied force. In this case, the torque experienced by the loop is:

$$\vec{T} = \vec{m} \times \vec{B}.$$

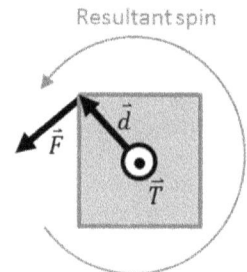

Torque (pointing out of the page) due to a force applied to an object on a fixed axis.

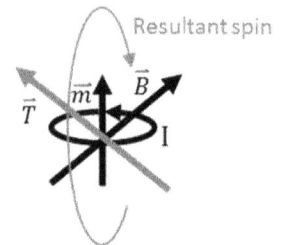

Torque resulting from a small current loop placed in a magnetic field.

Example

Find the torque experienced by a solenoid of radius a along the x-axis with a clockwise current in a magnetic field $\vec{B} = 5\hat{y} + 2\hat{z}$ mT.

$$\vec{m} = NIA\hat{n} = -NI\pi a^2 \hat{x}$$

$$\vec{B} = 5\hat{y} + 2\hat{z}$$

$$\vec{T} = \vec{m} \times \vec{B}$$

$$= NI\pi a^2 (2\hat{y} - 5\hat{z}) \quad \text{mN·m}$$

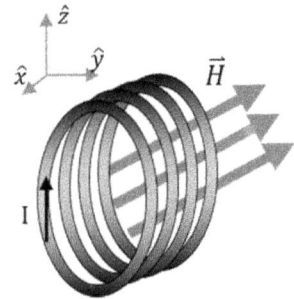

Solenoid in a magnetic field.

Example

Find the torque experienced by a wire loop at an angle α to a current carrying wire.

$$\vec{m} = NIA\hat{n} = I_1 \pi a^2 (-\sin\alpha\,\hat{r} + \cos\alpha\,\hat{z})$$

$$\vec{B_2} = \frac{\mu I_2}{2\pi r} \hat{\theta}$$

$$\vec{T} = \vec{m} \times \vec{B}$$

$$= \frac{\mu a^2 I_1 I_2}{2r} (-\cos\alpha\,\hat{r} - \sin\alpha\,\hat{z}) \quad \text{N·m}$$

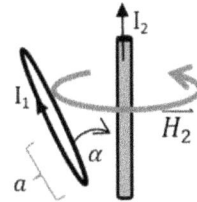

Wire loop next to a straight wire.

Part 2: Dynamics

Chapter 8: Equations for Dynamics

Faraday's Law:

Michael Faraday was convinced that if a current could produce a magnetic field, then a magnetic field should induce current. In 1832, he was finally able to prove this concept of **electromagnetic induction**. He found that a time-varying magnetic field near a wire produces an **electromotive force (emf)**, V_{emf} measured in Volts (V). The relation is called **Faraday's law**:

$$V_{emf} = \frac{-N\Phi}{dt} = -N\frac{d}{dt}\int \vec{B} \cdot d\vec{S} = \oint \vec{E} \cdot d\vec{L}.$$

There are three ways in which a magnetic field can induce current as detailed below.

1. **Transformer emf**: The force produced on a stationary loop in a time-varying magnetic field:

$$V_{tr,emf} = -N\int \frac{d\vec{B}}{dt} \cdot d\vec{S}.$$

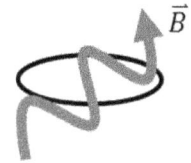

Transformer emf.

2. **Motional emf**: The force produced on a wire whose area is changing in the presence of a static magnetic field.

$$V_{m,emf} = -N\int \vec{B} \cdot \frac{d(d\vec{S})}{dt} = \oint (\vec{v} \times \vec{B}) \cdot d\vec{L}.$$

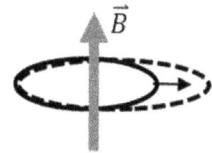

Motional emf.

3. **General emf**: The force produced when both the wire and magnetic field are changing.

$$V_{emf} = V_{tr,emf} + V_{m,emf}.$$

The current induced follows **Lentz's law** which states that the direction of the current acts to oppose the magnetic field that induced it. We'll consider a few examples in the next pages.

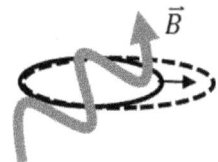

Most general case of emf.

> **Example**
>
> Find the relations between voltage and current in the two wires of the ideal (lossless) transformer.
>
> A **transformer** is an object that couples the current and voltage of one system to another with scaling.
>
> In the figure to the right, we see that an oscillatory voltage source (noted by the tilde in a circle) is connected to the first coil. This will produce a time-varying current which in turn will produce a flux carried by the dielectric that is the same for both coils (another example of mutual inductance).
>
> From Faraday's law, we have:
>
> $$V_1 = -N_1 \frac{d\Phi}{dt}; \quad V_2 = -N_2 \frac{d\Phi}{dt};$$
>
> Setting terms equal for the same time-varying flux:
>
> $$\frac{V_1}{V_2} = \frac{N_1}{N_2}$$
>
> Since the system is lossless,
>
> $$P_1 = V_1 I_1 = P_2 = V_2 I_2$$
>
> which gives:
>
> $$\frac{V_1}{V_2} = \frac{N_1}{N_2} = \frac{I_2}{I_1} \qquad \text{Ideal Transformer}$$
>
> The transformer is an example of transformer emf, hence the name.

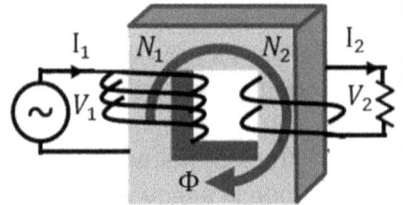

A transformer couples voltage and current with scaling through two wire loops.

> **Example**

Find the current produced by a wire moving across a U-shaped conductor at a velocity, $\vec{v} = v\hat{x}$ in a static magnetic field given by $\vec{B} = B\hat{z}$.

From Faraday's Law for motional emf:

$$V = \oint (\vec{v} \times \vec{B}) \cdot d\vec{L}$$

Since $\vec{v} = 0$ for all paths except the moving bar, the integral collapses to:

$$V = \int_0^l (v\hat{x} \times B\hat{z}) \cdot dy\,\hat{y} = -vBl$$

Also, we could say:

$$V = -\frac{d}{dt}\int_0^l \int_0^x B\hat{z} \cdot dx\,dy\,\hat{z} = -\frac{d}{dt}Bxl$$

$$= -\frac{d}{dt}Bvtl = -vBl$$

In the last step, we used the relation $v = x/t$.

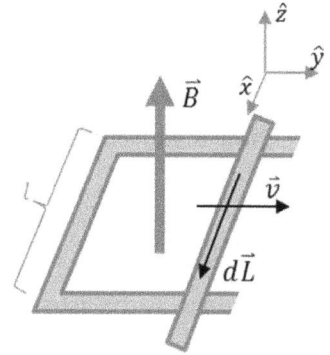

A changing wire shape in a steady magnetic field.

> **Example**

Find the current produced by a wire spinning with a speed ω in a constant magnetic field $\vec{B} = B\hat{z}$.

Again we have motional emf. The upper edge of the loop moves at a velocity of $\vec{v} = \frac{\omega d}{2}\hat{n}$, and the lower edge moves with an equal and opposite velocity. As $\hat{n}$ is always changing as the loop rotates:

$$\hat{n} \times \hat{z} = \sin \omega t\,\hat{x},$$

But, $d\vec{L} = dx\,\hat{x}$ always.

$$V = \int_0^l \frac{\omega d}{2} B \sin \omega t\, dx - \int_0^l \frac{-\omega d}{2} B \sin \omega t\, dx$$

$$= \omega dlB \sin \theta = \omega dlB \sin \omega t.$$

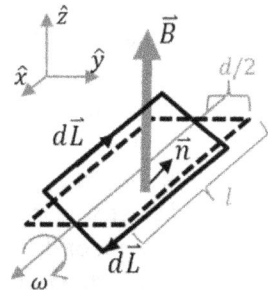

A rotating wire in a steady magnetic field.

Ampère's Law:

Ampère's law for dynamics is slightly different than in the static case:

$$\oint \vec{H} \cdot d\vec{L} = I = \int (\vec{J} + \frac{d\vec{D}}{dt}) \cdot d\vec{S}.$$

Now the total current is given by:

$$I = \frac{-dQ}{dt} = I_c + I_d = \int \left(\vec{J} + \frac{d\vec{D}}{dt}\right) \cdot d\vec{S}$$

where I_c is the usual conduction current, I_d is called the **displacement current**, and $\frac{d\vec{D}}{dt}$ is the **displacement current density**.

To see the significance of a displacement current density, consider a capacitor wired to a time-varying voltage source. We know from **Kirchhoff's current law** that the current has to be the same going in as going out of the capacitor:

$$\oint \vec{J} \cdot d\vec{S} = 0.$$

But, dielectrics don't conduct current – there aren't any free electrons. What does happen is that the dielectric's dipoles change polarity with the changing electric field. In response, the free charges on the conductor surfaces change, as well, thus transmitting the current as though the dielectric wasn't there. The changing dipole orientation, or displacement of the electrons relative to the nuclei, due to the changing electric field gives rise to the term displacement current. From the above results, we can deduce the **charge continuity equation**:

$$\nabla \cdot \vec{J} = \frac{-d\rho_v}{dt}.$$

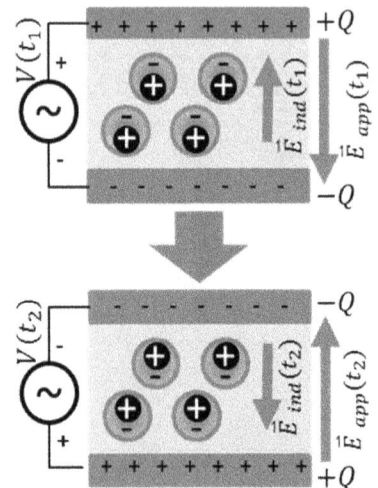

Changing electric field in a dielectric gives rise to a displacement current.

Find the displacement and conduction currents for a capacitor hooked up to a voltage source:

$$V = V_0 \sin \omega t \ \text{V}$$

We'll start by finding the displacement current from the electric field inside the capacitor:

$$V = -\int \vec{E}_{ind} \cdot d\vec{L} = -E_{ind,z} h = V_0 \sin \omega t$$

$$\vec{E}_{ind} = \frac{-V_0}{h} \sin \omega t \ \hat{z} \to \vec{D} = \varepsilon \vec{E} = \frac{-\varepsilon V_0}{h} \sin \omega t \ \hat{z}$$

$$I_d = \int_0^w \int_0^l \frac{d\vec{D}}{dt} \cdot dx dy \hat{z} = \frac{-\varepsilon \omega A V_0}{h} \cos \omega t \quad \text{Amps}$$

where $A = Area = wl$.

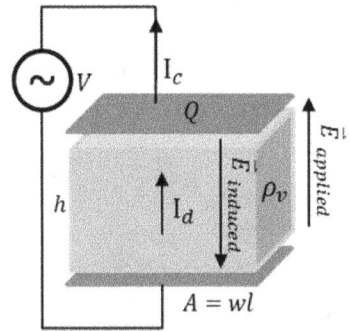

Conduction and displacement current in a capacitor.

Outside the capacitor, there is no displacement current, and the applied field is equal and opposite to the induced field, so we can write:

$$I_c = I - I_d = I = -\frac{\partial Q}{\partial t} = -\frac{\partial}{\partial t} \int \varepsilon \vec{E}_{app} \cdot d\vec{S}$$

$$= -\frac{\partial}{\partial t} \left(\frac{\varepsilon A V_0}{h} \sin \omega t \right) = \frac{-\varepsilon \omega A V_0}{h} \cos \omega t \quad \text{Amps}$$

The displacement current carries the charge across the capacitor as if the dielectric were not there. Kirchhoff's current law is also obeyed with current going into the top plate, I_d, equal to current going out of the plate, I_c.

Maxwell's Equations for Dynamics:

Now we can finally put together **Maxwell's equations for dynamics** including Faraday's law and Ampère's complete law. In **integral form** we have:

$$\oint \vec{E} \cdot d\vec{L} = \frac{-d}{dt} \int \vec{B} \cdot d\vec{S} \qquad \oint \vec{H} \cdot dL = \int \left(\vec{J} + \frac{d\vec{D}}{dt} \right) \cdot d\vec{S}$$

$$\oiint \vec{D} \cdot d\vec{S} = Q \qquad \oiint \vec{B} \cdot d\vec{S} = 0$$

With some mathematical derivation using the Stoke's and Divergence Theorems we get the **derivative, or point, form**:

$$\nabla \times \vec{E} = -\frac{d\vec{B}}{dt} \qquad \nabla \times \vec{H} = \vec{J} + \frac{d\vec{D}}{dt}$$

$$\nabla \cdot \vec{D} = \rho_v \qquad \nabla \cdot \vec{B} = 0$$

The **constituent relations** remain the same:

$$\vec{D} = \varepsilon \vec{E} \qquad \vec{J} = \sigma \vec{E} \qquad \vec{B} = \mu \vec{H}$$

Finally, the dielectric-dielectric **boundary conditions** remain unchanged and can be derived just as we did in the static case:

$$D_{n1} - D_{n2} = \rho_s \qquad \varepsilon_1 D_{t1} = \varepsilon_2 D_{t2}$$

$$\varepsilon_1 E_{n1} - \varepsilon_2 E_{n2} = \rho_s \qquad E_{t1} = E_{t2}$$

$$B_{n1} = B_{n2} \qquad \frac{B_{t1}}{\mu_1} = \frac{B_{t2}}{\mu_2}$$

$$\mu_1 H_{n1} = \mu_2 H_{n2} \qquad H_{t1} = H_{t2}$$

Retarded Potentials:

Before leaving this chapter, we consider one other topic: the electric and magnetic potentials for dynamics. In the static case we had:

$$\vec{E} = -\nabla V; \ \vec{B} = \nabla \times \vec{A}.$$

Now with Faraday's law we have:

$$\nabla \times \vec{E} = -\frac{d\vec{B}}{dt} = -\frac{d}{dt}\left(\nabla \times \vec{A}\right) = -\nabla \times \frac{d\vec{A}}{dt} \rightarrow$$

$$\nabla \times \left(\vec{E} + \frac{d\vec{A}}{dt}\right) = 0.$$

The last equation mimics the static case of Gauss's law, $\nabla \times \vec{E} = 0$, when we derived $\vec{E} = -\nabla V$. This implies:

$$\left(\vec{E} + \frac{d\vec{A}}{dt}\right) = -\nabla V \rightarrow \vec{E} = -\nabla V - \frac{d\vec{A}}{dt}.$$

Using the above result, we can derive Poisson's equations for dynamics:

$$\nabla \cdot \vec{D} = \nabla \cdot \varepsilon\vec{E} = \rho_v$$

$$\varepsilon\nabla \cdot \left(-\nabla V - \frac{d\vec{A}}{dt}\right) = -\varepsilon\left(\nabla^2 V + \frac{d}{dt}\nabla \cdot \vec{A}\right) = \rho_v$$

As we did before, we can define $\vec{A}$ any way we please as long as it doesn't violate Maxwell's equations. We make the following definition:

$$\nabla \cdot \vec{A} = -\mu\varepsilon\frac{dV}{dt}.$$

Now we have:

$$\rho_v = -\varepsilon\left(\nabla^2 V + \frac{d}{dt}\left(-\mu\varepsilon\frac{dV}{dt}\right)\right) \rightarrow$$

$$\nabla^2 V = \frac{-\rho_v}{\varepsilon} + \mu\varepsilon\frac{d^2 V}{dt^2}.$$

Similarly, we can derive the expression for the magnetic potential:

$$\nabla \times \vec{H} = \vec{J} + \frac{d\vec{D}}{dt}$$

$$\frac{1}{\mu}\nabla \times \nabla \times \vec{A} = \vec{J} + \varepsilon\frac{d\vec{E}}{dt} = \vec{J} - \varepsilon\frac{d}{dt}\left(\nabla V + \frac{d\vec{A}}{dt}\right)$$

$$\frac{1}{\mu}\left(\nabla(\nabla \cdot \vec{A}) - \nabla^2\vec{A}\right) = \frac{1}{\mu}\left(\nabla\left(-\mu\varepsilon\frac{dV}{dt}\right) - \nabla^2\vec{A}\right)$$

$$= \vec{J} - \varepsilon\frac{d}{dt}\left(\nabla V + \frac{d\vec{A}}{dt}\right) \rightarrow$$

$$\nabla^2\vec{A} = -\mu\vec{J} + \mu\varepsilon\frac{d^2\vec{A}}{dt^2}.$$

Now to address the term **retarded potential**: when in a time-varying electro-magnetic field, the effects of that field at a distance, r_o, from the source is time delayed by a factor of $\Delta t = r_o/v_p$ where v_p is the **propagation velocity**. The familiar forms for finding potential from a source are now:

Static Field	Dynamic Field
$V(r_o) = \frac{1}{4\pi\varepsilon}\int\frac{\rho_v(r_o)}{r_o}dv'$	$V(r_o, t) = \frac{1}{4\pi\varepsilon}\int\frac{\rho_v(r_o, t-\Delta t)}{r_o}dv'$
$\vec{A}(r_o) = \frac{\mu}{4\pi}\int\frac{\vec{J}(r_o)}{r_o}dv'$	$\vec{A}(r_o, t) = \frac{\mu}{4\pi}\int\frac{\vec{J}(r_o, t-\Delta t)}{r_o}dv'$

In other words, in using the expressions for the source, you must go back to the source state at time $t - \Delta t$.

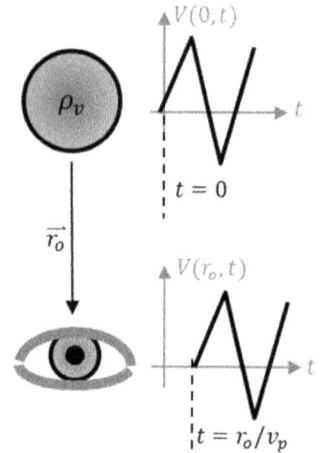

It takes time for forces and energy to travel to an observer.

Example

Find the potential at a point $(0,0,z)$ from a ring of line charge $\rho_L(t) = \rho_L \sin \omega t$ C/m with radius a.

We'll use the adaptation on Coulomb's law for line charge with:

$$r_o = \sqrt{a^2 + z^2}; \; dL' = a\partial\theta;$$

$$\rho_L(r_o, t) = \rho_L \sin\left(\omega\left(t - \frac{r_o}{v_p}\right)\right)$$

$$V = \int_0^{2\pi} \frac{\rho_L \sin\left(\omega\left(t - \frac{r_o}{v_p}\right)\right) a\partial\theta}{4\pi\varepsilon\sqrt{a^2+z^2}} = \frac{a\rho_L \sin\left(\omega\left(t - \frac{r_o}{v_p}\right)\right)}{2\varepsilon\sqrt{a^2+z^2}} \text{ V}$$

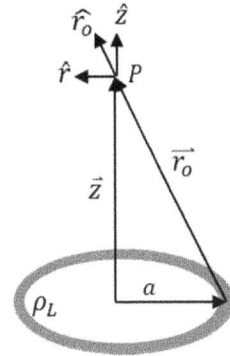

A ring of line charge.

Chapter 9: Electro-Magnetic Waves

An Introduction to Waves:

There is a special case of electro-magneto-dynamics where the electric and magnetic fields are time harmonic, i.e. they oscillate as sinusoids. The fields are called **electro-magnetic waves (EM waves)**, and they are perhaps the most useful fields you'll study here.

In this course, we'll consider the **plane wave**, a wave with constant phase fronts and uniform amplitude that is infinite in space and time. Since that (ahem) might be just a tad hard to imagine, let's instead consider a realistic portion of the wave. If you can picture a rippled potato chip or a Sun Chip™ traveling through space as shown to the right, you'll notice a few things:

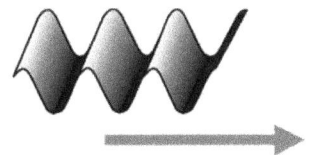

A portion of an EM plane wave.

- If you took a picture of the chip, you see it spatially vary as a sinusoid, just like in the figure.
- If you stood still as it passed by you and placed your hand on the top of (a very large scale version of) the chip, your hand would go up and down indicating a sinusoidal dependence on time.
- If you were to characterize this flying chip, you would call out the following:
  - The distance peak to crest, or amplitude.
  - The orientation of the chip – is it parallel to the ground, at an angle to the ground, or is it spinning like a football as it travels?
  - The direction in which it is traveling.
  - How fast it is traveling.
  - When it started traveling.
  - The distance between peaks, or wavelength. The wavelength is related to the speed of travel and the frequency of seeing a peak whizz by you if you stand still.

By the way, feel free to share all of this fun-filled insight at your next family barbeque.

In order to answer all of these questions about the plane waves whizzing by us, physicists came up with a way to describe them mathematically. A sample expression for the electric field plane wave is given below:

$$\vec{E}(x,y,z,t) = \vec{E}(x,y)\ e^{-jkz}\ e^{j(\omega t - \phi)}$$

$\omega = Frequency$
$\phi = Phase$

$k \propto {}^{\omega}/_{Speed(v_p)}$
$z = Direction$

$\vec{E}(x,y) = E_x\hat{x} + E_y\hat{y}$
Amplitude & Polarization

The one for the magnetic field is identical with E's replaced with H's. There are a couple of things that are different than you are used to in statics:

- First, there are exponential terms with imaginary numbers. Euler's identity gives us:

$$e^{j\theta} = \cos\theta + j\sin\theta.$$

Although we will use the exponential notation to better characterize the field's effects on other systems, in order to picture the field we'll just consider the real components.

- Second, the vector portion of the field, $\vec{E}(x,y)$, indicates amplitude and polarization (discussed in more detail in the next chapter) rather than direction. In other words, the vector indicates the magnitude and orientation in which the field oscillates rather than where it is going.

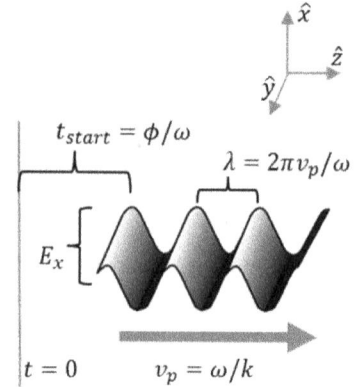

$t_{start} = \phi/\omega$

$\lambda = 2\pi v_p/\omega$

E_x

$t = 0$ $\quad$ $v_p = \omega/k$

Characteristics of an EM wave electric field. It is typical to make $\hat{z}$ the direction of travel. In this example, the electric field is given by $E_x\hat{x}e^{-jkz}e^{j(\omega t-\phi)}$.

A Note on Direction:

My background is in Electrical Engineering, so I use $e^{-jkz}e^{j\omega t}$ to indicate movement in the $+\hat{z}$ direction.

In Physics, however, the signs are switched so that $e^{jkz}e^{-j\omega t}$ indicates moving in the $+\hat{z}$ direction.

Also, it is possible to move in directions other than the z-axis. The wave $E_x\hat{x}e^{-j0.7k(y+z)}e^{j\omega t}$ moves at a 45° angle to the z-axis. More on this later.

In defining the electro-magnetic wave, we introduced some new terms:

Term	Symbol	Definition	Equations	Units
Frequency	ω, f	Number of crests in a unit of distance	$\omega = 2\pi f$	ω=rad/s, f=1/s=Hz
Wavelength	λ	Distance between crests	$\lambda = {v_p}/{f}$	m
Phase	ϕ	Indicates start or peak position	$\phi = \omega t_{start}$	rad
Phase Velocity	v_p	Speed of wave crests	$v_p = {1}/{\sqrt{\varepsilon\mu}}$	m/s

As you can see, these terms are related to each other. We define another term called the **propagation constant** or **wave number**, k, measured in 1/m, that captures these relationships:

$$Re\{k\} = \beta = \frac{2\pi n}{\lambda_0} = \frac{\omega}{v_p} = \omega\sqrt{\mu\varepsilon} = \sqrt{\varepsilon_r\mu_r}k_o = nk_o$$

where $n = \sqrt{\varepsilon_r\mu_r}$ is called the **index of refraction** or **refractive index**, k_o is the **propagation constant of free space**, and λ_0 is the wavelength in free space. The imaginary terms of k arise in lossy media as we'll discuss in Chapter 11. In the meantime, we can just use $Re\{k\} = k$.

Note that in a vacuum and free space, the phase velocity, or speed of the waves, is simply:

$$v_{p,free\ space} = \frac{1}{\sqrt{\varepsilon_0\mu_0}} = 3 \times 10^8\ \ m/s = c$$

where c is the speed of light. The equation for k implies that a wave traveling through a dielectric ($n > 1$) slows down and shortens in wavelength.

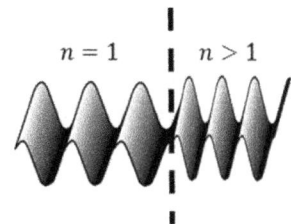

$n = 1$ $n > 1$

When a wave goes to a region of higher refractive index:
$$v_{p2} = \frac{c}{n};\ \lambda_2 = \frac{\lambda_0}{n}.$$

Maxwell's Equations for Waves:

Maxwell's Equations for waves become very simple. All of the $\frac{d}{dt}$'s become $j\omega$:

$$\oint \vec{E} \cdot d\vec{L} = -j\omega \int \vec{B} \cdot d\vec{S} \qquad \oint \vec{H} \cdot dL = \int(\vec{J} + j\omega\vec{D}) \cdot d\vec{S}$$

$$\oiint \vec{D} \cdot d\vec{S} = Q \qquad\qquad \oiint \vec{B} \cdot d\vec{S} = 0$$

$$\nabla \times \vec{E} = -j\omega\vec{B} \qquad\qquad \nabla \times \vec{H} = \vec{J} + j\omega\vec{D}$$

$$\nabla \cdot \vec{D} = \rho_v \qquad\qquad \nabla \cdot \vec{B} = 0$$

From the point form of these equations we can intuit that the vector components of the electric and magnetic fields are orthogonal to each other. In fact, the frequency, phase, speed, and direction remain the same between the fields, but they oscillate 90° from each other. This arrangement is characteristic of the **electro-magnetic wave (EM wave)**.

Example Find the magnetic field associated with the electric field given by: $\vec{E} = E_x\hat{x}\, e^{j\omega t}e^{-jkz}$ V/m

$$\nabla \times \vec{E} = \frac{\partial}{\partial z}\left(E_x e^{j\omega t}e^{-jkz}\right)\hat{y} = -j\omega\mu\vec{H}$$

$$-jk\, E_x e^{j\omega t}e^{-jkz}\hat{y} = -j\omega\mu\vec{H}$$

$$\vec{H} = \frac{k}{\omega\mu}E_x\, e^{j\omega t}e^{-jkz}\hat{y} \equiv H_y\hat{y}\, e^{j\omega t}e^{-jkz} \text{ A/m}$$

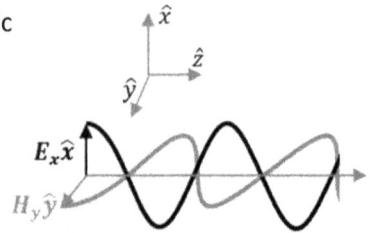

An example of an electro-magnetic wave.

The ratio of the magnetic field amplitude to the electric field amplitude is called the **intrinsic impedance** of the material, η, measured in Ohms:

$$\eta = \frac{|E|}{|H|} = \frac{\omega\mu}{k} = \frac{\omega\mu}{\omega\sqrt{\mu\varepsilon}} = \sqrt{\frac{\mu}{\varepsilon}} = \eta_0\sqrt{\frac{\mu_r}{\varepsilon_r}} \; ;$$

$$\eta_0 = \sqrt{\frac{\mu_0}{\varepsilon_0}} = 120\pi = 377\Omega .$$

The Wave Equation:

From Maxwell's equations, we can derive the **Wave equation**, also called the **Helmholtz equation**, for waves in a sourceless ($\rho_v = 0$; $\vec{J} = 0$) medium:

$$\nabla \times \vec{E} = -j\omega\mu\vec{H}$$

$$\nabla \times \nabla \times \vec{E} = -j\omega\mu\,\nabla \times \vec{H} = -j\omega\mu\,(j\omega\varepsilon\vec{E})$$

$$\nabla(\nabla \cdot \vec{E}) - \nabla^2\vec{E} = \omega^2\varepsilon\mu\vec{E} \quad (\nabla \cdot \vec{D} = 0)$$

$$\nabla^2\vec{E} + k^2\vec{E} = 0.$$

Similarly, we can find the wave equation for the magnetic field from $\nabla \times \vec{H} = j\omega\varepsilon\vec{E}$:

$$\nabla^2\vec{H} + k^2\vec{H} = 0.$$

The wave equation gives us the general form for the electric and magnetic fields. Writing it out explicitly:

$$\left(\frac{\partial^2}{\partial x^2} + \frac{\partial^2}{\partial y^2} + \frac{\partial^2}{\partial z^2}\right)\vec{E} = -k^2\vec{E}.$$

For a plane wave travelling in the $\hat{z}$ direction, the amplitude and polarization components (the x and y terms) are uniform. This leaves us with:

$$\frac{\partial^2}{\partial z^2}\vec{E} = -k^2\vec{E} \rightarrow \vec{E} = E_0 e^{-jkz}$$

which is the origin of our exponential $-jkz$ term. The wave equation plays an important role in optics, waveguides, and resonators.

Power & Energy:

The final topic of this chapter characterizes the power and energy in an electro-magnetic wave. As previously described, the electric and magnetic field pairs have the same frequency, phase, velocity, and direction. Their amplitudes differ by a factor of the intrinsic impedance, and they oscillate in directions 90° rotated from each other.

There is one other constraint forced upon the relationships in an electro-magnetic field. The **Poynting theorem** states that the power delivery (in W/m²) of an EM wave is:

$$\vec{\mathcal{P}} = \vec{E} \times \vec{H}; \quad \vec{\mathcal{P}}_{ave} = \frac{1}{2} Re\{\vec{E} \times \vec{H}^*\}$$

where the "*" indicates the complex conjugate.

The **Poynting vector**, $\vec{\mathcal{P}}$, not only gives the magnitude of the power density, but it also gives the direction in which the power flows. Because of the cross product, the power will always flow in the direction the wave travels and perpendicular to the direction of oscillation.

In addition, the Poynting theorem tells us that the electric and magnetic fields must obey the right hand rule: curling the fingers of your right hand from the direction of oscillation of $\vec{E}$ to the direction of oscillation of $\vec{H}$ must give the direction of travel.

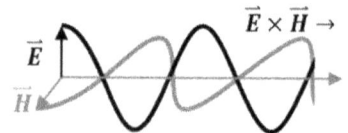

For an electro-magnetic wave, $\vec{E} \times \vec{H}$ must point in the direction of travel and power flow.

In terms of **potential energy** (in Joules), the equations for dynamics and waves look similar to those of the static field case except for an additional factor of ½ owing to time averaging:

$$W_e = \frac{1}{2} \int \varepsilon |\vec{E}|^2 \, dv.$$

$$W_m = \frac{1}{2} \int \mu |\vec{H}|^2 \, dv.$$

Find the average power density and energies associated with the electric field given by:

$$\vec{E} = E_x \hat{x}\, e^{j\omega t} e^{-jkz} \quad \text{V/m.}$$

We already found $\vec{H}$ in a prior example:

$$\nabla \times \vec{E} = \frac{\partial}{\partial z}\left(E_x\, e^{j\omega t}\, e^{-jkz}\right)\hat{y} = -j\omega\mu\vec{H}$$

$$\vec{H} = \frac{k}{\omega\mu}\, E_x\, e^{j\omega t}\, e^{-jkz}\, \hat{y} = \frac{E_x}{\eta}\hat{y}\, e^{j\omega t} e^{-jkz} \text{ A/m}$$

The average power is given by:

$$\vec{\mathcal{P}}_{ave} = \frac{1}{2}Re\{\vec{E} \times \vec{H}^*\}$$

$$= \frac{E_x^{\,2}}{2\eta}\hat{z}\left(e^{j\omega t}e^{-jkz}\right)\left(e^{-j\omega t}e^{jkz}\right) = \frac{E_x^{\,2}}{2\eta}\hat{z} \quad \text{W/m}^2$$

The average electric energy is (recall that $\left|\vec{E}\right|^2 = \vec{E} \cdot \vec{E}^*$):

$$W_e = \frac{1}{2}\int \varepsilon\left|\vec{E}\right|^2 dv = \frac{1}{2}\int \varepsilon E_x^{\,2}\, dv = \frac{\varepsilon v}{2}E_x^{\,2} \text{ J}$$

where v is volume. Similarly,

$$W_m = \frac{1}{2}\int \mu\left|\vec{H}\right|^2 dv = \frac{1}{2}\int \frac{\mu}{\eta^2}E_x^{\,2}\, dv = \frac{\mu v}{2\eta^2}E_x^{\,2} \text{ J}$$

With $\eta^2 = \mu/\varepsilon$ we get:

$$W_m = \frac{\varepsilon v}{2}E_x^{\,2} = W_e$$

Chapter 10: Polarization

Types of Polarization:

The topic of **polarization** can seem at times daunting. It can be difficult to picture and a bit abstract. So, let's work on getting a proper mental picture first. Polarization is a term that describes the direction of oscillation of a field – does it oscillate up and down, side to side, or does it turn as it travels through space?

There are three types of polarization – linear, circular, and elliptical. **Linear polarization** means that the fields oscillate along a straight line as they travel. We'll use the electric field for ease of illustration.

If we were to look head-on to a linearly polarized electric field, we would see this progression with time (the real part of $\vec{E}$):

$$\vec{E} = E_x\hat{x}\ e^{-jkz}\ e^{j\omega t}$$

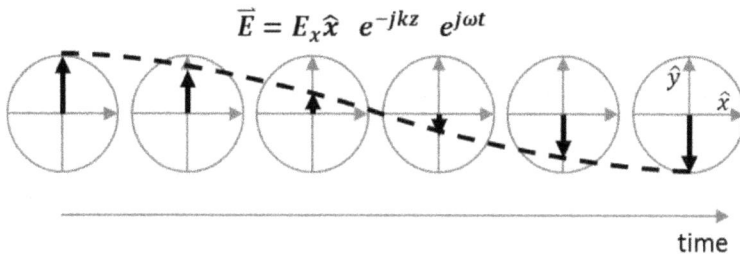

time

A linearly polarized electric field.

We might also see something that looks like this:

$$\vec{E} = (E_x\hat{x} + E_y\hat{y})e^{-jkz}\ e^{j\omega t}$$

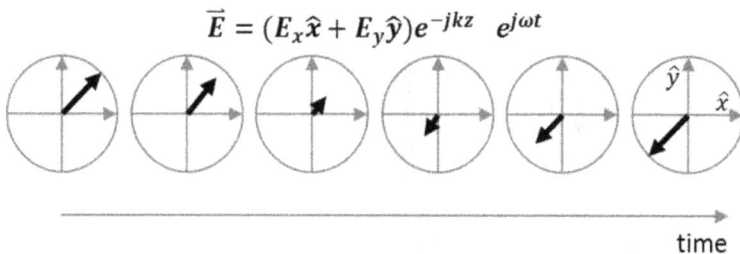

time

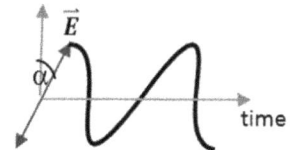

Another linearly polarized electric field oscillating at an angle to the page.

In both cases, the electric field is linearly polarized – it oscillates along a constant direction as it travels.

Circular polarization means that the fields are oscillating at a constant magnitude, but they are spinning as they do so.

$$\vec{E} = E_0(\hat{x} - j\hat{y})e^{-jkz} \; e^{j\omega t}$$

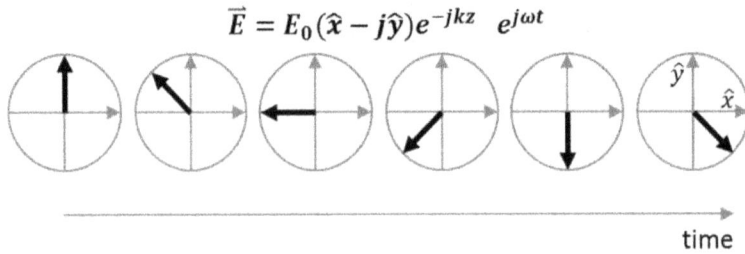

time

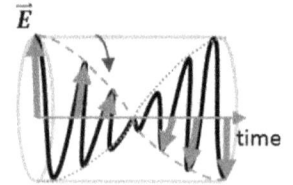

A (right hand) circularly polarized electric field.

For circular polarization, imagine the crests of the wave outlining a perfectly circular helix (or DNA molecule).

Elliptical polarization is the most general case where the fields are spinning, but the magnitude of oscillation is not necessarily equal on the $\hat{x}$ and $\hat{y}$ axes.

$$\vec{E} = (E_x\hat{x} + jE_y\hat{y})e^{-jkz} \; e^{j\omega t}$$

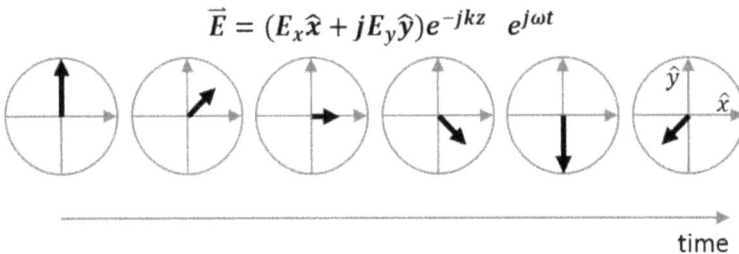

time

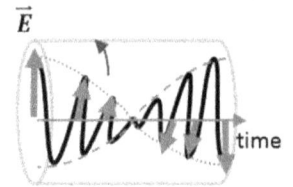

A (left hand) elliptically polarized electric field.

We can shorthand these time-valued vector representations as follows:

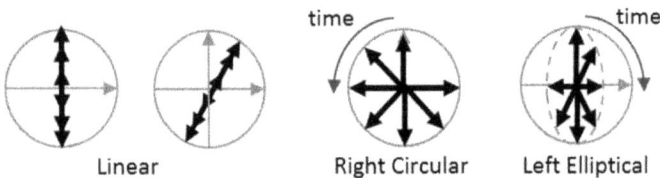

Linear   Right Circular   Left Elliptical

In all cases, we are viewing the wave as it comes toward us. The "right" and "left" nomenclature (for non-linear polarization) follow from curling the fingers as the vector amplitude rotates and having the thumb point in the direction of propagation.

Determining the Type of Polarization:

Determining the type and "handedness" of polarization is pretty easy once you have a good mental picture of what's going on. The first step is to determine linear, circular, or elliptical polarization.

For linear polarization, the distinction is very easy – the vector portion of $\vec{E}$ is real. That is, for propagation in the $\hat{z}$ direction:

$$\vec{E}(x,y) = E_x\hat{x} \pm E_y\hat{y} \quad \ni E_x, E_y \in \mathbb{R} \qquad \text{Linear}$$

Note: $\vec{E}(x,y) = jE_x\hat{x} \pm jE_y\hat{y}$ is also linearly polarized, though usually the imaginary term would be folded into the phase term as $e^{j(\omega t + \pi/2)}$ since $e^{j\pi/2} = j$.

For circular polarization, there is an imaginary component in the vector space, but the amplitudes of both the real and imaginary components are the same. For $\hat{z}$ propagation, that looks like:

$$\vec{E}(x,y) = E_o(\hat{x} \pm j\hat{y}) \quad \ni E_o \in \mathbb{R} \qquad \text{Circular}$$

Placing j on the $\hat{x}$ component rather than the $\hat{y}$ only means a phase shift in the $e^{j\omega t}$ term.

Finally, elliptical polarization is the hybrid of the two polarizations. Linear and circular polarization are actually special cases of elliptical polarization. For $\hat{z}$ propagation, the most general case looks like:

$$\vec{E}(x,y) = E_x\hat{x} \pm jE_y\hat{y} \quad \ni E_x, E_y \in \mathbb{R} \quad \text{Elliptical}$$

For the linearly polarized case, we are done. For the circular and elliptical cases, we have one more step to fully characterize the polarization of the wave.

The second step is to determine the "handedness" or direction of spin. For this, we need to construct diagrams similar to the shorthand time-valued vector representations of the preceding section. The good news is that we only have to take two time steps.

In order to resolve the wave's time progression, we plot the real part (the part we can see) as it travels toward us. We will plot the real part of the field at two convenient times:

1. t_0 where $e^{j(\omega t + \phi)} = 1$;
2. t_1 where $e^{j\left(\omega t + \phi + \frac{\pi}{2}\right)} = j$.

Example

Determine the polarization of
$$\vec{E} = E_0(\hat{x} - j\hat{y})e^{-jkz}e^{j\omt}$$
and determine the corresponding $\vec{H}$ field.

Amplitude is the same for both $\hat{x}$ and $\hat{y}$ with a factor of j on one component → Circular.

1. At time t_0, $Re\{\overrightarrow{E_{xy}}\} = E_0\hat{x}$;
2. At time t_1, $Re\{\overrightarrow{E_{xy}}\} = Re\{E_0(j\hat{x} - j^2\hat{y})\} = E_0\hat{y}$

With thumbs pointed toward us, the *right* fingers curl from time t_0 to time t_1 giving us right hand circular (RHC) polarization.

The magnetic field must obey Poynting's theorem by leading $\vec{E}$ by $\frac{\pi}{2}$:

$$\vec{H} = \frac{1}{\eta}E_0(\hat{x} - j\hat{y})e^{-jkz}e^{j\omega t}e^{\frac{j\pi}{2}}$$

$$= \frac{1}{\eta}E_0(j\hat{x} + \hat{y})e^{-jkz}e^{j\omega t}$$

$$\vec{P}_{ave} = \frac{1}{2}Re\{\vec{E} \times \overrightarrow{H^*}\} = \frac{1}{\eta}E_0{}^2\hat{z}$$

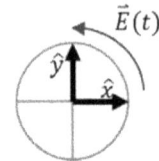

$\vec{E}(t)$

Electric field going from
$\vec{E}(t_0) \rightarrow \vec{E}(t_1)$.

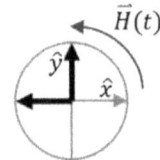

$\vec{H}(t)$

Magnetic field going
from $\vec{H}(t_0) \rightarrow \vec{H}(t_1)$.

Example

Determine the polarization of
$$\vec{E} = (E_x\hat{x} + jE_y\hat{y})e^{-jkz}e^{j\omega t}; \quad E_y = 2E_x$$
and determine the corresponding $\vec{H}$ field.

Amplitude is different for $\hat{x}$ and $\hat{y}$ with a factor of j on one component $\rightarrow$ Elliptical.

1. At time t_0, $Re\{\overrightarrow{E_{xy}}\} = E_x\hat{x}$;
2. At time t_1, $Re\{\overrightarrow{E_{xy}}\} = -E_y\hat{y} = -2E_x\hat{y}$

With thumbs pointed toward us, the *left* fingers curl from time t_0 to time t_1 giving us left hand elliptical (LHE) polarization.

The magnetic field must obey Poynting's theorem by right hand rule no matter what "handedness" the electric field obeys. The magnetic field itself will also be left-handed elliptical, but at each time, t, the Poynting vector will point in the $+\hat{z}$ direction dictated by our e^{-jkz} term. So, this time, $\vec{H}$ lags $\vec{E}$ by $\pi/2$, so we multiply $\vec{E}$ by $-j/\eta$ to get $\vec{H}$:

$$\vec{H} = \frac{1}{\eta}\left(E_x\hat{x} + jE_y\hat{y}\right)e^{-jkz}e^{j\omega t}e^{\frac{-j\pi}{2}}$$

$$= \frac{1}{\eta}\left(-jE_x\hat{x} + E_y\hat{y}\right)e^{-jkz}e^{j\omega t}$$

$$\vec{\mathcal{P}}_{ave} = \frac{1}{2}Re\{\vec{E} \times \overrightarrow{H^*}\} = 2\frac{1}{\eta}E_x^2\hat{z}$$

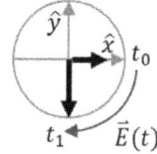

Electric field going from $\vec{E}(t_0) \rightarrow \vec{E}(t_1)$.

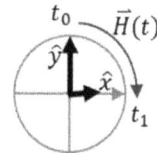

Magnetic field going from $\vec{H}(t_0) \rightarrow \vec{H}(t_1)$. Note that at each time, $\vec{E} \times \vec{H}$ points in the direction of propagation, $+\hat{z}$, or out of the page.

In summary:
- Plot $Re\{\overrightarrow{E_{xy}}(t_0)\} = Re\{1 \times \overrightarrow{E_{xy}}\}$;
- Plot $Re\{\overrightarrow{E_{xy}}(t_1)\} = Re\{j \times \overrightarrow{E_{xy}}\}$;
- Check sign on e^{-jkz} term to verify propagation direction ($e^{j\omega t}e^{+jkz}$ goes in the $-\hat{z}$ direction);
- $\overrightarrow{H_{xy}} = (\pm j/\eta)\overrightarrow{E_{xy}}$ to obey Poynting's theorem;
- Check $Re\{\vec{E} \times \overrightarrow{H^*}\}$ points toward propagation.

Chapter 11: Boundaries at Normal Incidence

Normal Incidence at a Single Boundary:

For ease of notation and readability, we'll be dropping the $e^{j(\omega t+\phi)}$ terms for this chapter. They are still there, and we know that in going from free space to a medium with n>1,

$$v_{p2} = \frac{c}{n}; \quad \lambda_2 = \frac{\lambda_0}{n}.$$

So, let's consider what happens to the other characteristics that make up our fields as they pass from one medium to another. We'll introduce four new terms to aid analysis:

- **Plane of incidence**: The plane that holds both the direction of propagation $(k\hat{r})$ and the boundary between the two materials.

- **Reflection coefficient**, Γ: The percentage of the incident field that is reflected back into medium 1.

$$\Gamma = \frac{|\vec{E}_r|}{|\vec{E}_i|} = \frac{\eta_2-\eta_1}{\eta_2+\eta_1}.$$

- **Transmission coefficient**, τ: The percentage of the incident field that is transmitted to medium 2.

$$\tau = \frac{|\vec{E}_t|}{|\vec{E}_i|} = \frac{2\eta_2}{\eta_2+\eta_1} = 1 + \Gamma.$$

- **Standing wave ratio**, SWR: The addition of the incident and reflected fields cause interference and the presence of a standing wave characterized by:

$$SWR = \frac{|\vec{E}_{1,max}|}{|\vec{E}_{1,min}|} = \frac{1+|\Gamma|}{1-|\Gamma|}.$$

Note: The terms for Γ and τ can be derived by equating tangential components at the boundary and solving:

$$E_i + \Gamma E_i = \tau E_i; \quad \frac{E_i}{\eta_1} + \frac{\Gamma E_i}{\eta_1} = \frac{\tau E_i}{\eta_2}$$

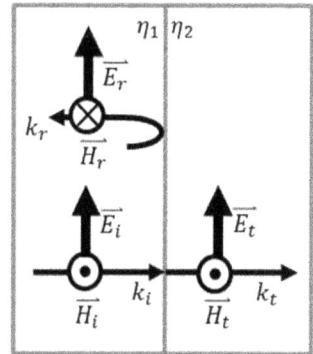

Geometry for normal incidence at a boundary.

Here we drew $\vec{H}$ normal to the page. We could have switched the roles of $\vec{E}$ and $\vec{H}$ which is equivalent and just requires rotating the plane of incidence by 90°.

Note on Γ and τ:

The power reflected is $|\Gamma|^2$.

The power transmitted is NOT $|\tau|^2$.
The power transmitted is $1 - |\Gamma|^2$.

Develop expressions for the incident, reflected, and transmitted left circularly polarized electric and magnetic waves from air to glass at normal incidence. Glass has properties: $\mu = \mu_0$; $\sqrt{\varepsilon_r} = 1.5$.

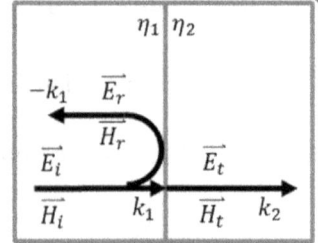

Geometry example problem.

For LHC, we write the incident wave as:

$$\overrightarrow{E_i} = E_0(\hat{x} + j\hat{y})e^{-jk_1z}e^{j\omega t}$$

$$\overrightarrow{H_i} = \frac{-j}{\eta_0}\overrightarrow{E_i} = \frac{E_0}{\eta_0}(-j\hat{x} + \hat{y})e^{-jk_1z}e^{j\omega t}$$

The transmitted wave will also be LHC travelling in the $+\hat{z}$ direction, but in medium 2 with an amplitude modified by τ :

$$\tau = \frac{2\eta_2}{\eta_2+\eta_1} = \frac{2\eta_0/1.5}{(1+\frac{1}{1.5})\eta_0} = \frac{4}{5}$$

$$\overrightarrow{E_t} = \frac{4}{5}E_0(\hat{x} + j\hat{y})e^{-jk_2z}e^{j\omega t}$$

$$\overrightarrow{H_t} = \frac{-j}{\eta_2}\overrightarrow{E_t} = \frac{4E_0}{5(\frac{\eta_0}{1.5})}(-j\hat{x} + \hat{y})e^{-jk_2z}e^{j\omega t}$$

$$= \frac{6E_0}{5\eta_0}(-j\hat{x} + \hat{y})e^{-jk_2z}e^{j\omega t}$$

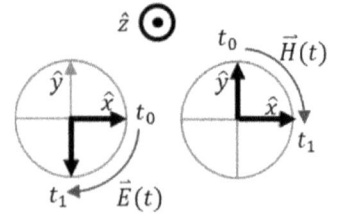

Checking polarization: When $\hat{z}$ points out of the page, $(\hat{x} + j\hat{y})$ and $(-j\hat{x} + \hat{y})$ are LHC and follow Poynting's theorem.

The reflected wave is LHC but travelling in the $-\hat{z}$ direction with an amplitude modified by Γ :

$$\Gamma = \frac{\eta_2-\eta_1}{\eta_2+\eta_1} = \frac{\frac{1}{1.5}-1}{\frac{1}{1.5}+1} = \frac{-1}{5};$$

$|\Gamma|^2 = 4\%$ power is reflected

$$\overrightarrow{E_r} = \frac{-1}{5}E_0(\hat{x} + j\hat{y})e^{+jk_1z}e^{j\omega t}$$

$$\overrightarrow{H_r} = \frac{+j}{\eta_0}\overrightarrow{E_r} = \frac{-E_0}{5\eta_0}(j\hat{x} - \hat{y})e^{+jk_1z}e^{j\omega t}$$

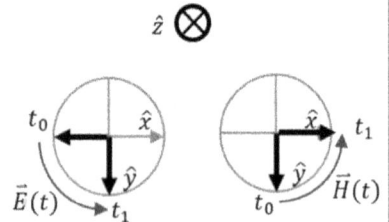

Checking polarization: When $\hat{z}$ points into the page, $(-\hat{x} - j\hat{y})$ and $(-j\hat{x} + \hat{y})$ are LHC and follow Poynting's theorem.

Normal Incidence at Multiple Boundaries:

With multiple boundaries, the situation becomes a little more complex. Consider two boundaries with three materials to start. Material 1 interacts with material 2, so it doesn't know much about material 3 except what results in material 2 because of material 3. If we can find an effective or **input impedance**, η_{in} (in Ohms), to fully characterize the system of materials 2 and 3, then we can evaluate the reflectance at the first border with:

$$\Gamma = \frac{|\vec{E_r}|}{|\vec{E_i}|} = \frac{\eta_{in}-\eta_1}{\eta_{in}+\eta_1}.$$

Recalling that $\eta = \frac{|\vec{E}|}{|\vec{H}|}$, and analyzing what is going on in medium 2 at $z = -l$ (which is what material 1 cares about):

$$\eta_{in} = \frac{E_2}{H_2}\Big|_{z=-l} = \frac{E_{2i}(e^{-jkz}+\Gamma_2 e^{+jkz})}{\frac{E_{2i}}{\eta_2}(e^{-jkz}-\Gamma_2 e^{+jkz})}\Big|_{z=-l}; \quad \Gamma_2 = \frac{\eta_3-\eta_2}{\eta_3+\eta_2};$$

$$\vdots$$

$$\eta_{in} = \frac{\eta_3 \cos k_2 l + j\eta_2 \sin k_2 l}{\eta_2 \cos k_2 l + j\eta_3 \sin k_2 l}; \quad k_2 = \frac{2\pi n_2}{\lambda_0}$$

For more than two boundaries, we repeat the same process moving against the direction of propagation (starting from the right and moving left in our example figure). In each step:

$$\Gamma_{n-1} = \frac{\eta_{in,n}-\eta_{in,n-1}}{\eta_{in,n}+\eta_{in,n-1}};$$

Important examples exist in the real world when $\Gamma = -1$ (total reflectance) or $\Gamma = 0$ (total transmission). As a laser scientist that needs to reflect and transmit light of a particular wavelength, I'm very concerned about coatings (material 2) on glass substrates (material 3) that totally reflect (like mirrors) or totally or partially transmit (like windows or laser output couplers).

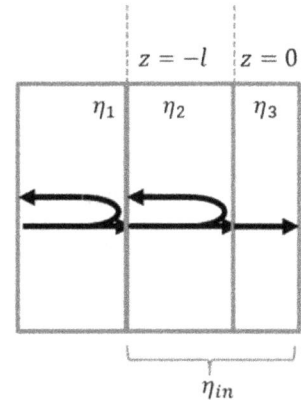

Geometry for normal incidence at two boundaries. The goal is replace the last two materials with an effective impedance.

Example

Develop a $\lambda/4$ coating that will allow green light (~550 nm) to fully transmit into glass from air.

As our coating will have a length of $\lambda/4$, η_{in} is:

$$\eta_{in} = \frac{\eta_3 \cos k_2 l + j\eta_2 \sin k_2 l}{\eta_2 \cos k_2 l + j\eta_3 \sin k_2 l} = \frac{\eta_2{}^2}{\eta_3}$$

We want $\Gamma = 0$ which requires $\eta_{in} = \eta_1 \rightarrow$

$$\eta_2 = \sqrt{\eta_1 \eta_3} \rightarrow n_2 = \sqrt{n_1 n_2} = \sqrt{1.5} = 1.22$$

So, to transmit green light, we want a $\frac{550}{4} = 137$ nm thick coating of a material with index 1.22.

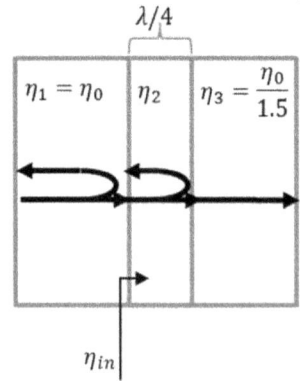

Coating sandwiched between air and glass.

Example

Determine the reflection from 3 and 6 sets of alternating $\lambda/4$ coatings as shown.

Again, our $\lambda/4$ coatings simplify η_{in}:

$$\eta_{in,6} = \frac{\eta_b{}^2}{\eta_g}; \qquad\qquad \eta_{in,5} = \frac{\eta_a{}^2}{\eta_{in,6}} = \frac{\eta_a{}^2 \eta_g}{\eta_b{}^2};$$

$$\eta_{in,4} = \frac{\eta_b{}^2}{\eta_{in,5}} = \frac{\eta_b{}^4}{\eta_a{}^2 \eta_g}; \quad \eta_{in,3} = \frac{\eta_a{}^2}{\eta_{in,4}} = \frac{\eta_a{}^4 \eta_g}{\eta_b{}^4};$$

$$\eta_{in,2} = \frac{\eta_b{}^2}{\eta_{in,3}} = \frac{\eta_b{}^6}{\eta_a{}^4 \eta_g}; \quad \eta_{in,1} = \frac{\eta_a{}^2}{\eta_{in,2}} = \frac{\eta_a{}^6 \eta_g}{\eta_b{}^6};$$

$$\frac{\eta_a{}^6 \eta_g}{\eta_b{}^6} = \left(\frac{n_b}{n_a}\right)^6 \frac{n_0}{n_g} = \left(\frac{1.22}{2.0}\right)^6 \frac{n_0}{1.5} = 0.034\, \eta_0;$$

$$\Gamma = \frac{\eta_{in,1} - \eta_1}{\eta_{in,1} + \eta_1} = \frac{0.034 - 1}{1.034} = -0.934; \ 87\% \text{ reflected.}$$

When there are 6 sets of coatings:

$$\eta_{in,1} = \frac{\eta_a{}^{12} \eta_g}{\eta_b{}^{12}} = \left(\frac{1.22}{2.0}\right)^{12} \frac{n_0}{1.5} = 0.0018\, \eta_0$$

$$\Gamma = \frac{\eta_{in,1} - \eta_1}{\eta_{in,1} + \eta_1} = \frac{0.0018 - 1}{1.0018} = -0.996; \ 99\% \text{ reflected.}$$

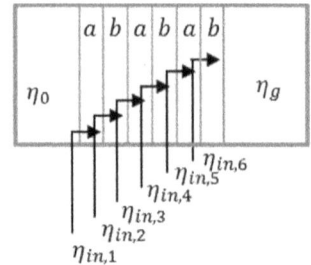

$$\eta_a = \frac{\eta_0}{2.0}; \eta_b = \frac{\eta_0}{1.22}; \eta_g = \frac{\eta_0}{1.5}$$

Geometry for high reflector – 3 sets of coatings shown.

Propagation & Attenuation:

In the real world, materials are not ideal dielectrics with no loss, nor are they ideal conductors with infinite conductivity. We'll define the permittivity in a way that will help us capture both the dielectric and conductor nature of a material. We start with:

$$\nabla \times \vec{H} = \vec{J} + j\omega\vec{D} = (\sigma + j\omega\varepsilon')\vec{E} = j\omega(\varepsilon' - j\frac{\sigma}{\omega}) \equiv j\omega\varepsilon\vec{E}.$$

In the above equation, we've renamed our old permittivity, ε', and we've assigned a new quantity to the variable ε:

$$\varepsilon = \varepsilon' - j\frac{\sigma}{\omega} \equiv \varepsilon' - j\varepsilon''; \quad \varepsilon'' = \frac{\sigma}{\omega}.$$

Now we can finally talk about the real and imaginary parts of the wavenumber, $k = \beta - j\alpha$:

$$jk = j\omega\sqrt{\mu\varepsilon} = j\omega\sqrt{\mu(\varepsilon' - j\varepsilon'')} = j\omega\sqrt{\mu\varepsilon'(1 - j\left(\frac{\varepsilon''}{\varepsilon'}\right)}$$
$$\equiv \alpha + j\beta$$

$$\alpha = \omega\sqrt{\frac{\mu\varepsilon'}{2}\left(\sqrt{1 + \left(\frac{\varepsilon''}{\varepsilon'}\right)^2} - 1\right)} \rightarrow \textbf{attenuation}$$

$$\beta = \omega\sqrt{\frac{\mu\varepsilon'}{2}\left(\sqrt{1 + \left(\frac{\varepsilon''}{\varepsilon'}\right)^2} + 1\right)} = \frac{2\pi n}{\lambda_0} \rightarrow \textbf{propagation}$$

$$e^{-jkz} = e^{-\alpha z}e^{-j\beta z}$$

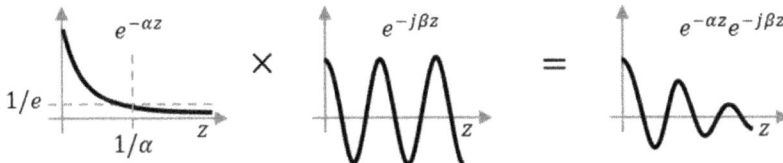

Both the **propagation constant**, β, and the **attenuation constant**, α, are measured in 1/m. Luckily, most materials come close to the ideal case, and we can shorten the cumbersome equations for α and β with some approximations:

For a dielectric: $\varepsilon' \gg \varepsilon''$:

$$\alpha \approx \omega \sqrt{\frac{\mu\varepsilon'}{2} \left(\sqrt{1 + (0)^2} - 1 \right)} = 0$$

$$\beta \approx \omega \sqrt{\frac{\mu\varepsilon'}{2} \left(\sqrt{1 + (0)^2} + 1 \right)} = \omega\sqrt{\mu\varepsilon'}$$

$$k = \beta - j\alpha \approx \beta$$

$$\eta = \sqrt{\frac{\mu}{\varepsilon}} = \sqrt{\left(\frac{\mu}{\varepsilon'}\right)\left(\frac{1}{1 - j^{\varepsilon''}/_{\varepsilon'}}\right)} \approx \sqrt{\frac{\mu}{\varepsilon'}}$$

These results are similar to what we've been using thus far.

For a conductor: $\varepsilon'' \gg \varepsilon'$:

$$\alpha = \omega \sqrt{\frac{\mu\varepsilon''}{2} \left(\sqrt{1 + \left(\frac{\varepsilon'}{\varepsilon''}\right)^2} - \frac{\varepsilon'}{\varepsilon''} \right)} \approx \omega\sqrt{\frac{\mu\varepsilon''}{2}} = \sqrt{\frac{\sigma\omega\mu}{2}}$$

$$\beta = \omega \sqrt{\frac{\mu\varepsilon''}{2} \left(\sqrt{1 + \left(\frac{\varepsilon'}{\varepsilon''}\right)^2} + \frac{\varepsilon'}{\varepsilon''} \right)} \approx \omega\sqrt{\frac{\mu\varepsilon''}{2}} = \sqrt{\frac{\sigma\omega\mu}{2}}$$

$$k = \beta - j\alpha \approx \sqrt{\frac{\sigma\omega\mu}{2}}(1 - j)$$

$$\eta = \sqrt{\frac{\mu}{\varepsilon}} = \sqrt{\left(\frac{\mu}{\varepsilon''}\right)\left(\frac{1}{\varepsilon'/_{\varepsilon''} - j}\right)} \approx \sqrt{\frac{\mu}{\varepsilon''}} = \sqrt{\frac{\omega\mu}{2\sigma}}(1 + j)$$

The above results are summarized on the next page.

Dielectric $\varepsilon' \gg \varepsilon''$	$\alpha \approx 0$	$\beta \approx \omega\sqrt{\mu\varepsilon'}$	$k \approx \beta$	$\eta \approx \sqrt{\dfrac{\mu}{\varepsilon'}}$
Conductor $\varepsilon'' \gg \varepsilon'$	$\alpha \approx \sqrt{\dfrac{\sigma\omega\mu}{2}}$	$\beta \approx \sqrt{\dfrac{\sigma\omega\mu}{2}}$	$k \approx \sqrt{\dfrac{\sigma\omega\mu}{2}}(1-j)$	$\eta \approx \sqrt{\dfrac{\omega\mu}{2\sigma}}(1+j)$

Because the wave attenuates ($\alpha \neq 0$) in a conductor, we can talk about a **skin depth**, δ (in meters) given by:

$$\delta = \sqrt{\frac{2}{\sigma\omega\mu}} = \frac{1}{\sqrt{\pi f \sigma \mu}}.$$

The skin depth is the distance a wave travels in a medium before being attenuated to $1/e$ of its original value.

Example

Find τ, δ, and the distance needed to drop to 10% amplitude for a linearly polarized electric field oscillating at 10 kHz incident from air to sea water ($\mu = \mu_0$; $\sqrt{\varepsilon_r} = 1.33$; $\sigma = 4.8$).

At $f = 10^4$, $\dfrac{\varepsilon''}{\varepsilon'} = \dfrac{\sigma}{\omega\varepsilon'} = \dfrac{4.8}{(2\pi f)1.33^2(\varepsilon_0)} = 4.9 \times 10^6$

Sea water is definitely a good conductor, so we use

$$\eta_2 = \sqrt{\left(\frac{\omega\mu}{2\sigma}\right)}(1+j) = 0.09\angle\pi/4$$

$$\tau = \frac{2\eta_2}{\eta_2+\eta_1} \approx \frac{0.18}{377} = 0.00048.$$

$$\alpha = \beta = \sqrt{\frac{\sigma\omega\mu}{2}} = 0.435$$

$$\delta = \frac{1}{\alpha} = 2.3 \text{ m.}$$

Linearly polarized electric field incident on sea water.

Distance to 10% of amplitude is given by:

$$e^{-\alpha z} = 0.1 \rightarrow z = 5.3 \text{ m.}$$

Chapter 12: Boundaries at Oblique Incidence

Oblique Incidence:

Before considering what happens to a linearly polarized wave when it hits a surface at an oblique angle, we need to develop expressions for waves that travel in a direction other than the z-axis.

Again, we'll consider a plane of incidence that contains the direction of travel and the boundary. In this case, it does matter which field, $\vec{E}$ or $\vec{H}$, is chosen to point into or out of the page. The two cases are defined by the electric field.

- If the electric field is perpendicular to the plane of incidence (comes out of the page), then we have **perpendicular polarization**, or **s-polarization**. The s stands for the German word for perpendicular which is *senkrecht*.
- If the electric field is parallel to the plane of incidence, then we have **parallel polarization**, or **p-polarization**. The p stands for the German word for parallel which is, oddly enough, *parallel*.

Electric and magnetic fields shown for s-polarization.

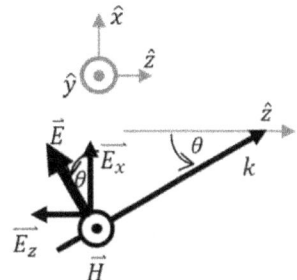

For an s-polarized wave traveling at an angle θ to the z-axis, our kr term becomes:

$$kr = k(x \sin \theta + z \cos \theta).$$

The polarization term for $\vec{E}$ is simply $E_0 \hat{y}$, so we have:

$$\vec{E} = E_0 \hat{y}\, e^{-jk(x \sin \theta + z \cos \theta)}.$$

Electric and magnetic fields shown for p-polarization.

For the p-polarized wave, we have the same kr term, but the polarization term becomes:

$$\vec{E} = E_0(\cos \theta\, \hat{x} - \sin \theta\, \hat{z}).$$

The final expression for the p-polarized electric field is:

$$\vec{E} = E_0(\cos \theta\, \hat{x} - \sin \theta\, \hat{z})e^{-jk(x \sin \theta + z \cos \theta)}.$$

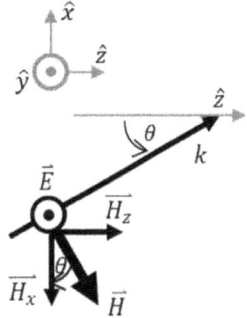

Perpendicular Polarization:

The terms for all of the fields shown for the s-polarized case is (get ready for some serious bookkeeping):

$$\vec{E_i} = E_i \hat{y} \, e^{-jk_1(x \sin \theta_i + z \cos \theta_i)}$$

$$\vec{E_r} = \Gamma E_i \hat{y} \, e^{-jk_1(x \sin \theta_r - z \cos \theta_r)}$$

$$\vec{E_t} = \tau E_i \hat{y} \, e^{-jk_2(x \sin \theta_t + z \cos \theta_t)}$$

$$\vec{H_i} = \frac{E_i}{\eta_1} (-\cos \theta_i \, \hat{x} + \sin \theta_i \, \hat{z}) e^{-jk_1(x \sin \theta_i + z \cos \theta_i)}$$

$$\vec{H_r} = \Gamma \frac{E_i}{\eta_1} (\cos \theta_r \, \hat{x} + \sin \theta_r \, \hat{z}) e^{-jk_1(x \sin \theta_r - z \cos \theta_r)}$$

$$\vec{H_t} = \tau \frac{E_i}{\eta_2} (-\cos \theta_t \, \hat{x} + \sin \theta_t \, \hat{z}) e^{-jk_2(x \sin \theta_t + z \cos \theta_t)}$$

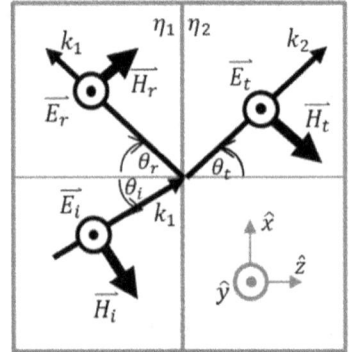

Electric and magnetic fields shown for s-polarization hitting a boundary at an oblique angle. Notice that $\vec{E} \times \vec{H}^*$ always points in the direction of propagation .

At the boundary ($z = 0$), we must match all of the phases (k terms) for continuity. This is called the **phase matching condition** which gives us:

$$k_1 x \sin \theta_i = k_1 x \sin \theta_r = k_2 x \sin \theta_t.$$

The first two quantities give us **Snell's Law of Reflection**:

$$\theta_i = \theta_r \equiv \theta_1.$$

The first and third quantities give us **Snell's Law of Refraction**:

$$k_1 \sin \theta_i = \frac{2\pi n_1}{\lambda_0} \sin \theta_i = k_2 \sin \theta_t = \frac{2\pi n_2}{\lambda_0} \sin \theta_t$$

$$n_1 \sin \theta_1 = n_2 \sin \theta_2 ; \quad \theta_t \equiv \theta_2.$$

These laws hold for the p-polarized case, as well, and will help simplify some of the notation.

There are a few more conditions which have to be met at the boundary. One, recall that the tangential components of the electric field must be equal on either side of $z = 0$:

$$\left(\overrightarrow{E_i} \cdot \hat{y} + \overrightarrow{E_r} \cdot \hat{y}\right)\Big|_{z=0} = \overrightarrow{E_t} \cdot \hat{y}\Big|_{z=0}$$

$$E_i \, e^{-jk_1(x \sin \theta_1)} + \Gamma E_i \, e^{-jk_1(x \sin \theta_1)} = \tau E_i \, e^{-jk_1(x \sin \theta_1)}$$

$$1 + \Gamma = \tau$$

In the second step, we used Snell's Law of Refraction on the last term. Just as the tangential components of $\vec{E}$ must be equal on the boundary, so do those of the magnetic field:

$$\left(\overrightarrow{H_i} \cdot \hat{x} + \overrightarrow{H_r} \cdot \hat{x}\right)\Big|_{z=0} = \overrightarrow{H_t} \cdot \hat{x}\Big|_{z=0}$$

$$\frac{E_i}{\eta_1}(-\cos \theta_1)e^{-jk_1(x \sin \theta_1)} + \Gamma \frac{E_i}{\eta_1}(\cos \theta_1)e^{-jk_1(x \sin \theta_1)}$$
$$= \tau \frac{E_i}{\eta_1}(-\cos \theta_2)e^{-jk_1(x \sin \theta_1)}$$

$$\frac{\cos \theta_1}{\eta_1}(-1 + \Gamma) = \frac{\cos \theta_2}{\eta_2}(-1 - \Gamma)$$

$$\Gamma = \frac{\eta_2 \sec \theta_2 - \eta_1 \sec \theta_1}{\eta_2 \sec \theta_2 + \eta_1 \sec \theta_1} \equiv \frac{\eta'_2 - \eta'_1}{\eta'_2 + \eta'_1}$$

We can find a similar term for τ though we won't derive it here. The final result is that for s-polarization, we can use the same equations as we did for normal incidence with a slight modification:

$$1 + \Gamma = \tau; \quad \Gamma = \frac{\eta'_2 - \eta'_1}{\eta'_2 + \eta'_1}; \quad \tau = \frac{2\eta'_2}{\eta'_2 + \eta'_1};$$

$$\eta'_i = \eta_i \sec \theta_i \quad \text{s-polarization}$$

Parallel Polarization:

The terms for all of the fields shown for the p-polarized case is:

$$\vec{E_i} = E_i \left(\cos\theta_1\,\hat{x} - \sin\theta_1\,\hat{z}\right)e^{-jk_1(x\sin\theta_1 + z\cos\theta_1)}$$

$$\vec{E_r} = \Gamma E_i\left(\cos\theta_1\,\hat{x} + \sin\theta_1\,\hat{z}\right)e^{-jk_1(x\sin\theta_1 - z\cos\theta_1)}$$

$$\vec{E_t} = \tau E_i\left(\cos\theta_2\,\hat{x} - \sin\theta_2\,\hat{z}\right)e^{-jk_2(x\sin\theta_2 + z\cos\theta_2)}$$

$$\vec{H_i} = \frac{E_i}{\eta_1}\hat{y}\,e^{-jk_1(x\sin\theta_1 + z\cos\theta_1)}$$

$$\vec{H_r} = -\Gamma\frac{E_i}{\eta_1}\hat{y}\,e^{-jk_1(x\sin\theta_1 - z\cos\theta_1)}$$

$$\vec{H_t} = \tau\frac{E_i}{\eta_2}\hat{y}\,e^{-jk_2(x\sin\theta_2 + z\cos\theta_2)}$$

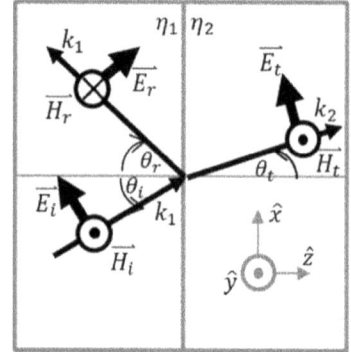

Electric and magnetic fields shown for p-polarization hitting a boundary at an oblique angle. Notice that $\vec{E} \times \vec{H}^*$ always points in the direction of propagation .

We've already applied Snell's laws and the angle notation for media 1 & 2.

Making sure the tangential components of the electric and magnetic field are equal at the boundary, and after a whole lot of math, we get:

$$1 + \Gamma = \tau; \quad \Gamma = \frac{\eta'_2 - \eta'_1}{\eta'_2 + \eta'_1}; \quad \tau = \frac{2\eta'_2}{\eta'_2 + \eta'_1};$$

$$\eta'_i = \eta_i \cos\theta_i \quad \text{p-polarization}$$

Example

Find expressions for the incident, reflected, and transmitted waves of a right-hand circularly polarized wave incident at 45° on an air-glass ($\sqrt{\varepsilon_r} = 1.5$) boundary.

To handle the polarization, we'll express the fields in terms of component vectors that are orthogonal to themselves and the direction of propagation:

$$\hat{s} = \hat{y}; \quad \hat{p} = \cos\theta\,\hat{x} - \sin\theta\,\hat{z}$$

The incident wave is straightforward, but tedious, based on combining our terms for RHC polarization and oblique incidence.

Component vectors for this example.

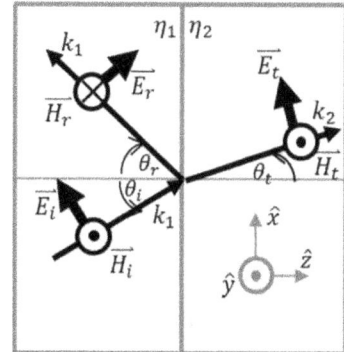

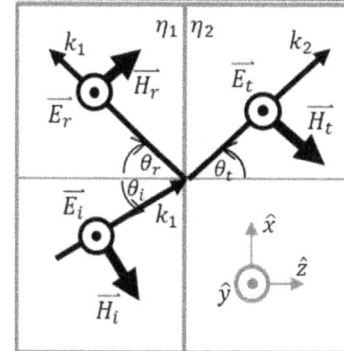

$$\vec{E_i} = E_i[\hat{p} - j\hat{s}]\, e^{-jk_1(x\sin\theta_1 + z\cos\theta_1)}$$
$$= E_i[(\cos\theta_1\,\hat{x} - \sin\theta_1\,\hat{z}) - j\hat{y})]\, e^{-jk_1\frac{\sqrt{2}}{2}(x+z)}$$
$$= \frac{\sqrt{2}}{2} E_i[(\hat{x} - \hat{z}) - j\hat{y})]\, e^{-jk_1\frac{\sqrt{2}}{2}(x+z)}$$

$$\vec{H_i} = \frac{E_i}{\eta_0}(j\hat{p} + \hat{s})e^{-jk_1\frac{\sqrt{2}}{2}(x+z)}$$
$$= \frac{\sqrt{2}}{2\eta_0} E_i[j(\hat{x} - \hat{z}) + \hat{y})]\, e^{-jk_1\frac{\sqrt{2}}{2}(x+z)}$$

To find Γ and τ, we use $\eta_2 = \frac{\eta_0}{1.5} = \frac{\eta_1}{1.5}$ and:

$$\theta_2 = \sin^{-1}\left(\frac{n_1}{n_2}\sin\theta_1\right) = \frac{\sqrt{2}}{2\cdot1.5} = 0.47$$

$$\cos\theta_1 = \frac{\sqrt{2}}{2} = 0.71; \quad \cos\theta_2 = 0.89;$$
$$\sec\theta_1 = \frac{2}{\sqrt{2}} = 1.41; \quad \sec\theta_2 = 1.12;$$

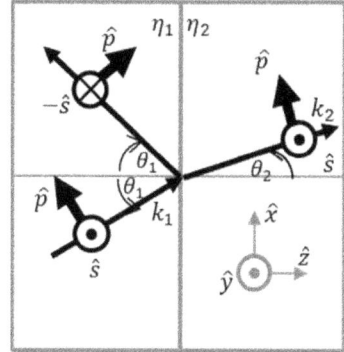

Circular polarization can be split into its p and s linear components.

So, we have:

Ex. cont.

$$\Gamma_p = \frac{\frac{0.89}{1.5}-0.71}{\frac{0.89}{1.5}+0.71} = -0.09; \quad \Gamma_s = \frac{\frac{1.12}{1.5}-1.41}{\frac{1.12}{1.5}+1.41} = -0.31;$$

$$\tau_p = 1 + \Gamma_p = 0.91; \qquad \tau_s = 1 + \Gamma_s = 0.69;$$

The transmitted wave is easiest, so we'll do that first:

$$\overrightarrow{E_t} = E_i(\tau_p\hat{p} - j\tau_s\hat{s})e^{-jk_2(x\sin\theta_2 + z\cos\theta_2)}$$

$$= E_i[0.91(0.89\hat{x} - 0.45\,\hat{z}) - j0.69\hat{y}]e^{-jk_2(0.45x+0.89z)}$$

$$= E_i[(0.81\hat{x} - 0.41\,\hat{z}) - j0.69\hat{y}]e^{-jk_2(0.45x+0.89z)}$$

$$\overrightarrow{H_t} = \frac{E_i}{\eta_2}(j\tau_p\hat{p} + \tau_s\hat{s})e^{-jk_2(x\sin\theta_2 + z\cos\theta_2)}$$

$$= \frac{E_i}{\eta_2}[j(0.81\hat{x} - 0.41\,\hat{z}) + 0.69\hat{y}]e^{-jk_2(0.45x+0.89z)}$$

For the reflected wave, we have to remember to obey Poynting's theorem. This time $\hat{p} = \cos\theta\,\hat{x} + \sin\theta\,\hat{z}$:

$$\overrightarrow{E_r} = E_i[-\Gamma_p\hat{p} - j\Gamma_s\hat{s}]e^{-jk_1(x\sin\theta_1 - z\cos\theta_1)}$$

$$= E_i[0.09(0.7)(\hat{x} + \hat{z}) + j0.31\hat{y}]e^{-jk_10.7(x-z)}$$

$$= E_i[0.06(\hat{x} + \hat{z}) + j0.31\hat{y}]e^{-jk_10.7(x-z)}$$

$$\overrightarrow{H_r} = \frac{E_i}{\eta_0}(-j\Gamma_p\hat{p} + \Gamma_s\hat{s})e^{-jk_10.7(x-z)}$$

$$= \frac{E_i}{\eta_0}[j0.06(\hat{z} + \hat{x}) - 0.31\hat{y}]e^{-jk_10.7(x-z)}$$

The Critical Angle:

Let's look again at Snell's Law of Refraction:

$$n_1 \sin \theta_1 = n_2 \sin \theta_2 .$$

When $n_1 < n_2$, the beam is refracted *toward* the z-axis. When $n_1 > n_2$, the beam is refracted *away from* the z-axis. In the latter case, one can imagine that as θ_2 increases faster than θ_1, there will come a point where $\theta_2 \to 90°$. That happens when:

$$\sin \theta_1 = \frac{n_2}{n_1} \sin \frac{\pi}{2} \quad \to \quad \theta_c = \sin^{-1} \frac{n_2}{n_1}$$

where θ_c is called the **critical angle**. Once θ_1 goes past the critical angle, the wave cannot pass into medium 2. Some aspect of the wave may travel along the boundary, but generally speaking, the wave is reflected.

The critical angle is particularly important in fiber optics. An optical fiber is made up of a glass core (about the diameter of a hair) surrounded by a glass cladding with slightly lower refractive index than that of the core. The cladding serves to protect the core and make sure the boundary stays pristine. By launching light at a slight angle relative to the axis of the fiber, it hits the core-cladding boundary at a steep angle relative to the normal. This steep angle is well past the critical angle causing what is termed as **total internal reflection** (TIR).

TIR is responsible for guiding the light hundreds of kilometers down the fiber core, but the fiber must be relatively straight. When the fiber is bent at too small a curvature, light hits the core-cladding boundary at an angle less than the critical angle causing leakage to the cladding. The light eventually couples completely out of the fiber due to impurities and imperfections at the cladding outer boundary.

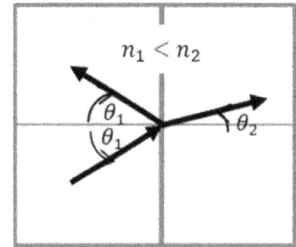
Reflection and refraction when $n_1 < n_2$.

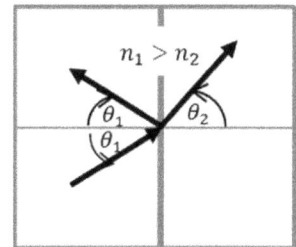
Reflection and refraction when $n_1 > n_2$.

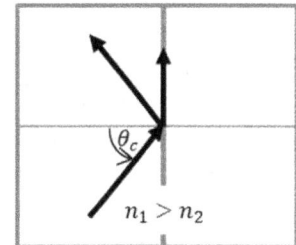
Reflection when $\theta_1 = \theta_c$ for $n_1 > n_2$.

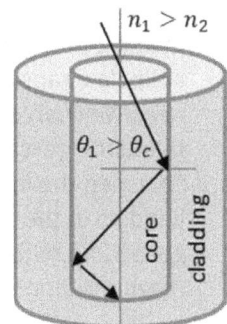
TIR in optical fiber.

Brewster's Angle:

Another special angle of incidence occurs when there is no reflection at the boundary. For the s-polarized case, this never occurs in nonmagnetic materials ($\mu \neq \mu_0$). However, in the p-polarized case we get:

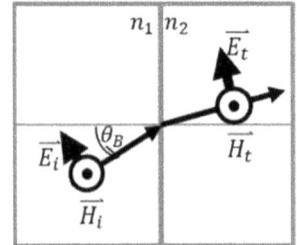

Transmission when $\theta_1 = \theta_B$ for p-polarization.

$$\Gamma = 0 = \frac{\eta_2 \cos\theta_2 - \eta_1 \cos\theta_1}{\eta_2 \cos\theta_2 + \eta_1 \cos\theta_1} \rightarrow \eta_2 \cos\theta_2 = \eta_1 \cos\theta_1$$

$$\cos^2\theta_1 = 1 - \sin^2\theta_1 = \left(\frac{n_1}{n_2}\right)^2 (1 - \sin^2\theta_2)$$

$$1 - \sin^2\theta_1 = \left(\frac{n_1}{n_2}\right)^2 \left(1 - \left(\frac{n_1}{n_2}\right)^2 \sin^2\theta_1\right)$$

And, after more mathematical manipulation, we get:

$$\sin\theta_B = n_2\sqrt{\frac{1}{n_1^2 + n_2^2}}$$

Here, we've let $\theta_1 \equiv \theta_B$, **Brewster's angle**, at which none of the incident wave is reflected for p-polarization. If mixed polarizations are incident at Brewster's angle, then the material acts as a **polarizer** – as all of the p-polarized component is transmitted, the only field that is reflected is s-polarized.

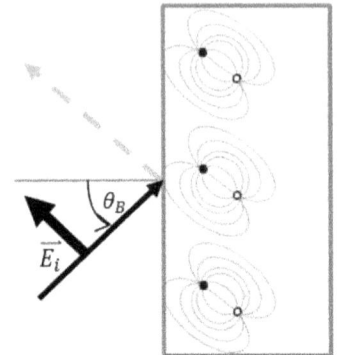

The electric field sets up dipoles that align to the reflected angle when Brewster's angle is reached.

Brewster's angle is caused by the impingent oscillating electric field which sets up oscillating electric dipoles in media 2 at the boundary. Recall that the resulting $\vec{E}$ field of a dipole wraps around from the positive charge to the negative charge. It does not emit purely along the axis of the dipole (except in the space right between the charges). So, when the incident angle is just right, the dipoles are lined up in the direction where the reflected beam should go, and no radiation occurs. In a non-magnetic material, no magnetic dipoles exist, so Brewster's angle does not apply.

Chapter 13: Transmission Lines

Transmission Line Characteristics:

Transmission lines are made up of two conductors separated by a dielectric that carry current and voltage to and from a generator (or source) and a load. They generally have inductance and resistance on the conductors, and capacitance and conductance in the dielectric. A **lossless transmission line** is a special case with no resistance or conductance.

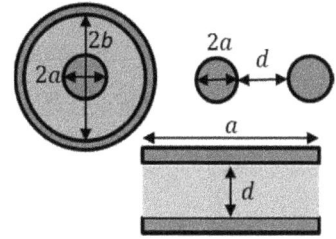

Cross-sections of some types of transmission lines.

	Coax	2-Wire	Parallel Plate	Units
L	$\frac{\mu}{2\pi}\ln\frac{b}{a}$	$\frac{\mu}{\pi}\ln\left[\frac{d}{2a}+\sqrt{\left(\frac{d}{2a}\right)^2-1}\right]$	$\frac{\mu d}{a}$	H/m
R*	$\sqrt{\frac{f\mu}{4\pi\sigma}}\left(\frac{1}{a}+\frac{1}{b}\right)$	$\sqrt{\frac{f\mu}{\pi\sigma}}\left(\frac{1}{a}\right)$	$\sqrt{\frac{4\pi f\mu}{\sigma}}\left(\frac{1}{a}\right)$	Ω/m
C	$\frac{2\pi\varepsilon}{\ln(b/a)}$	$\frac{\pi\varepsilon}{\ln\left[\frac{d}{2a}+\sqrt{\left(\frac{d}{2a}\right)^2-1}\right]}$	$\frac{\varepsilon a}{d}$	F/m
G	$\frac{2\pi\sigma}{\ln(b/a)}$	$\frac{\pi\sigma}{\ln\left[\frac{d}{2a}+\sqrt{\left(\frac{d}{2a}\right)^2-1}\right]}$	$\frac{\sigma a}{d}$	S/m

Adapted from Ulaby, Fundamentals of Applied Electromagnetics, 2004.
*μ, σ are properties of conductor for R, elsewise they refer to dielectric.

The **lumped element model** of a transmission line places the L, R, C, and G characteristics on a per length basis in a circuit diagram as shown. Using Kirchhoff's voltage and current laws, we can find the **transmission line equations**:

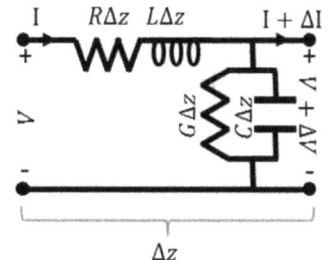

$$V = I(R\Delta z + j\omega L\Delta z) + V + \Delta V \rightarrow \frac{dV}{dz} = -(R+j\omega L)I;$$

$$I \approx V(G\Delta z + j\omega C\Delta z) + I + \Delta I \rightarrow \frac{dI}{dz} = -(G+j\omega C)V;$$

Lumped element model of transmission lines.

where the voltage across the parallel circuit elements is $V + \Delta z \approx V$.

We can take the second derivative of the transmission line equations to get something that looks very similar to the wave equation:

$$\frac{\partial^2 V}{\partial z^2} = -(R + j\omega L)\frac{\partial I}{\partial z} = (R + j\omega L)(G + j\omega C)V \equiv \gamma^2 V;$$

$$\frac{\partial^2 I}{\partial z^2} = -(G + j\omega C)\frac{\partial V}{\partial z} = (R + j\omega L)(G + j\omega C)I \equiv \gamma^2 I;$$

$$\frac{\partial^2 V}{\partial z^2} - \gamma^2 V = 0; \qquad \frac{\partial^2 I}{\partial z^2} - \gamma^2 I = 0;$$

$$\gamma^2 = (R + j\omega L)(G + j\omega C).$$

where γ [1/m] is called the **complex propagation constant** of the transmission line. Similar to jk, γ can be expressed as $\alpha + j\beta$.

Since the voltage and current waves can go in both directions, the solutions to the transmission line wave equations are:

$$V = V_\rightarrow e^{-\gamma z} + V_\leftarrow e^{\gamma z}; \quad I = I_\rightarrow e^{-\gamma z} + I_\leftarrow e^{\gamma z}.$$

We can use these expressions, the transmission line equations, and the definition of γ to define a **characteristic impedance**, Z_0 [Ω], for the line. The characteristic impedance gives the ratio of voltage to current going in *one direction*. We'll derive the forward direction, and give the result for the backward direction:

$$V_\rightarrow, I_\rightarrow + V_\leftarrow, I_\leftarrow$$

$$Z_0$$

$$\frac{\partial V_\rightarrow}{\partial z} = -\gamma V_\rightarrow e^{-\gamma z} = -(R + j\omega L)I_\rightarrow e^{-\gamma z};$$

$$Z_0 = \frac{V_\rightarrow}{I_\rightarrow} = \frac{R + j\omega L}{\gamma} = \sqrt{\frac{R + j\omega L}{G + j\omega C}} = -\frac{V_\leftarrow}{I_\leftarrow}.$$

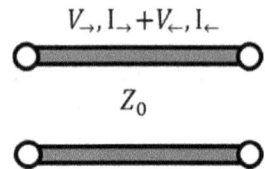

Voltage and current can travel in both directions of the transmission line. The characteristic impedance relates voltage to current in one direction.

Reflections & Standing Waves:

The **load impedance**, Z_L [Ω], is the ratio of the total voltage to total current going through the load. We can also define a **load admittance**, Y_L [$1/\Omega$], which is simply the inverse of the load impedance.

Generator, or source, circuit connected to a load via a transmission line.

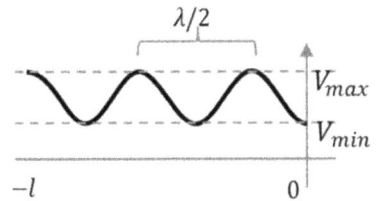

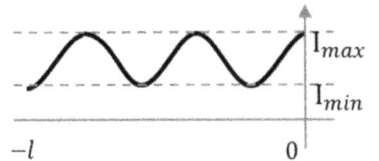

$$Z_L = \frac{V_L}{I_L} = \frac{V_\rightarrow + V_\leftarrow}{Z_0(V_\rightarrow - V_\leftarrow)}; \quad Y_L = \frac{1}{Z_L}.$$

The **reflection coefficient**, Γ, is the percentage of signal that is returned back on the line. It is defined in the same way we defined the reflectance at a boundary:

$$\Gamma = \frac{V_\leftarrow}{V_\rightarrow} = \frac{Z_L - Z_0}{Z_L + Z_0} = -\frac{I_\leftarrow}{I_\rightarrow}.$$

With Γ defined, we can rewrite the expressions for V and I on the line with $V_\rightarrow \equiv V_0$, $I_\rightarrow \equiv I_0$, and assuming a lossless line, $\gamma = j\beta$:

$$V(z) = V_0(e^{-j\beta z} + \Gamma e^{j\beta z});$$

$$I(z) = I_0\left(e^{-j\beta z} - \Gamma e^{j\beta z}\right) = \frac{V_0}{Z_0}\left(e^{-j\beta z} - \Gamma e^{j\beta z}\right).$$

Because of reflections from the load, interference between the forward and backward propagating waves on the line give rise to standing waves and a **standing wave ratio**, SWR:

Standing waves occur when lines are not matched to the load.

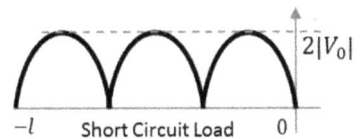

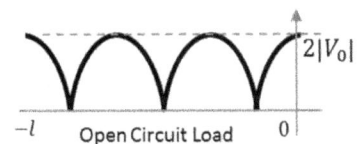

$$SWR = \frac{|V_{max}|}{|V_{min}|} = \frac{1 + |\Gamma|}{1 - |\Gamma|}.$$

Short Circuit Load

The SWR can vary from a value of 1, a **matched line** with no load reflections, to ∞, a completely mismatched load. If reflections do occur, two properties are always obeyed:

1. Maxima and minima are separated by $\lambda/2$, and
2. Voltage maxima correspond to current minima and vice versa.

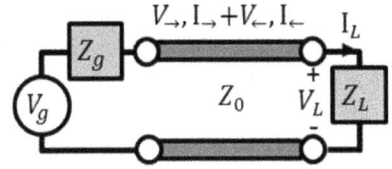

Open Circuit Load

Special cases of standing waves.

Find the standing wave ratio of a coaxial cable operating at 1 GHz connected to a 2.5 pF capacitor load. The cable has the following characteristics:

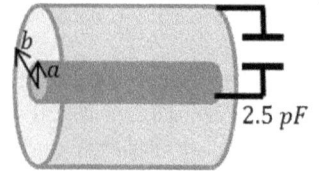

$$\mu = \mu_0; \quad \varepsilon = 2.25\varepsilon_0; \quad \sigma_c = 10^8; \quad \sigma_d = 10^{-4};$$

$$a = 1\text{mm}; \quad b = 5\text{mm}.$$

Coaxial transmission line connected to a capacitive load.

We'll find SWR by finding Γ for which we'll need Z_0. For a coaxial transmission line with $\omega = 2\pi f = 6.28 \times 10^9$, we have:

$$L = \frac{\mu}{2\pi}\ln\frac{b}{a} = 3.2 \times 10^{-7}; \quad j\omega L = j2010$$

$$R = \sqrt{\frac{f\mu}{4\pi\sigma_c}}\left(\frac{1}{a}+\frac{1}{b}\right) = 1.2 \times 10^{-3}$$

$$C = \frac{2\pi\varepsilon}{\ln(b/a)} = 7.78 \times 10^{-11}; \quad j\omega C = j0.5$$

$$G = \frac{2\pi\sigma_d}{\ln(b/a)} = 3.9 \times 10^{-4}$$

$$Z_0 = \sqrt{\frac{R+j\omega L}{G+j\omega C}} \approx \sqrt{\frac{j2010}{j0.5}} = 63.4 \ \Omega$$

$$Z_L = \frac{1}{j\omega C} \approx -j63.4$$

$$\Gamma = \frac{63.4(-j-1)}{63.4(-j+1)} = \frac{1\angle 5\pi/4}{1\angle -\pi/4} = 1\angle\frac{3\pi}{2} = -j$$

$$|\Gamma| = 1 \rightarrow SWR = \infty$$

The load is perfectly mismatched to the line causing a reflected wave that is 90° out of phase with the incident wave. The result is complete interference.

Input Impedance:

The final impedance that is important to a transmission line is its **input impedance**, Z_{in} [Ω], which is the ratio of total voltage to total current anywhere on the line. Because the voltage and current vary as a function of position, Z_{in} is also a function of z:

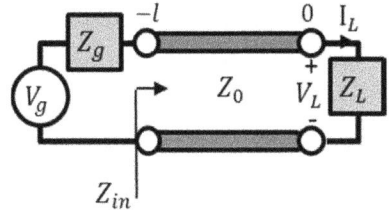

$$Z_{in}(z) = \frac{V(z)}{I(z)} = \frac{V_0(e^{-j\beta z}+\Gamma e^{j\beta z})}{V_0(e^{-j\beta z}-\Gamma e^{j\beta z})}Z_0 = Z_0\frac{1+\Gamma e^{j2\beta z}}{1-\Gamma e^{j2\beta z}}.$$

Input impedance looking at the start of the transmission line.

The input impedance at the beginning of the transmission line, $z = -l$, is:

$$Z_{in}(-l) \equiv Z_{in} = Z_0\frac{Z_L+jZ_0\tan\beta l}{Z_0+jZ_L\tan\beta l}$$

which allows us to represent the transmission line and load with an equivalent circuit as shown to the right.

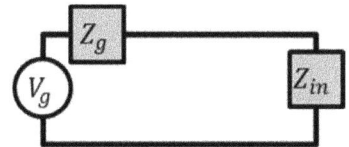

Using the input impedance to construct an equivalent circuit for the transmission line.

- A short or open circuit load gives:

$$Z_L = 0 \rightarrow Z_{in}{}^{sc} = jZ_0\tan\beta l;$$
$$Z_L = \infty \rightarrow Z_{in}{}^{oc} = -jZ_0\cot\beta l.$$

Note that $Z_0 = \sqrt{Z_{in}{}^{sc}Z_{in}{}^{oc}}$ and $\tan\beta l = \sqrt{\frac{-Z_{in}{}^{sc}}{Z_{in}{}^{oc}}}$ which gives us a way to make measurements on the line with open and short circuited loads to find the characteristic impedance and propagation constant of the line.

- If the line is matched, $Z_{in} = Z_0 = Z_L$, and there are no reflections.

- If the line has a length of multiples of $\lambda/2$, we have $Z_{in} = Z_L$ which is equivalent to the line not being there.

- If the line has a length of $\lambda/4$ plus multiples of $\lambda/2$, then $Z_{in} = Z_0^2/Z_L$. This type of line is called a **transformer** because it eliminates mismatches and reflections at its input. It is often used to tie a load to an existing line.

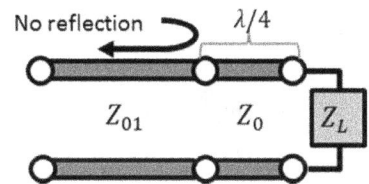

A $\lambda/4$ transformer.

One other way of eliminating reflections is to use a matching **shunt stub** on the line. The shunt stub is a shorted transmission line of a certain length placed in parallel with the original line a certain distance from the load. The final configuration can be represented by the equivalent circuit shown. The input impedance looking into the load at the stub location becomes:

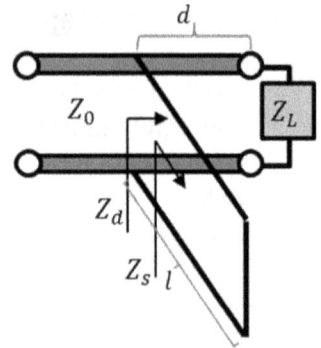

$$\frac{1}{Z_{in}} = \frac{1}{Z_s} + \frac{1}{Z_d} \quad \text{or equivalently} \quad Y_{in} = Y_s + Y_d$$

where $Y_i = 1/Z_i$. We'll revisit shunt stubs in the next chapter when we use Smith Charts.

A shunt stub with Z_d the input impedance from the stub to the load, and Z_s the input impedance of the stub..

As our last topic in this section, we have the power delivered by the line. The instantaneous power delivered to the load is given by:

$$P_{\rightarrow}(t) = \Re\{V_{\rightarrow}(z)e^{j(\omega t + \phi)}\}\Re\{I_{\rightarrow}(z)e^{j(\omega t + \phi)}\}$$

$$P_{\rightarrow}(t) = \frac{|V_{\rightarrow}(z)|^2}{Z_0}\cos^2(\omega t + \phi).$$

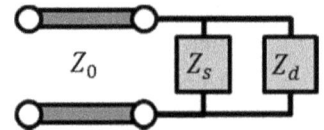

Equivalent circuit diagram for shunt stub.

Similarly,

$$P_{\leftarrow}(t) = -|\Gamma|^2 \frac{|V_{\leftarrow}(z)|^2}{Z_0}\cos^2(\omega t + \phi + \phi_{refl}).$$

The time-averaged power is found by:

$$P_{ave,\rightarrow} = \frac{1}{\tau}\int_0^\tau P_{\rightarrow}(t)\,dt = \frac{\omega}{2\pi}\int_0^{2\pi/\omega} P_{\rightarrow}(t)\,dt = \frac{1}{2}P_{\rightarrow}(t);$$

$$P_{ave,\leftarrow} = \frac{\omega}{2\pi}\int_0^{2\pi/\omega} P_{\leftarrow}(t)\,dt = \frac{1}{2}P_{\leftarrow}(t);$$

$$P_{ave} = P_{ave,\rightarrow} + P_{ave,\leftarrow} = \frac{|V_0|^2}{2Z_0}[1 - |\Gamma|^2] = \frac{1}{2}\Re\{\vec{V}\cdot\vec{I}^*\}.$$

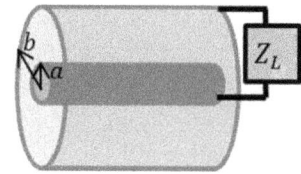

Coaxial transmission line
connected to various loads.

Example Find the average power delivered by the 1 GHz coaxial cable of the prior example connected to a 2.5 pF capacitor, a 50 Ω resistor, and the capacitor and resister in series.

We determined from the last example that
$$Z_0 = 63.4 \ \Omega.$$

The capacitive load has an impedance of
$$Z_L = \frac{1}{j\omega C} \approx -j63.4 \ \Omega$$

which gave us $|\Gamma| = 1$.

$$P_{ave} = \frac{|V_0|^2}{2Z_0}[1 - |\Gamma|^2] = 0 \ W$$

For the purely resistive load,
$$Z_L = 50 \ \Omega$$

$$\Gamma = \frac{50-63.4}{50+63.4} = -0.12$$

$$P_{ave} = \frac{|V_0|^2}{2Z_0}[1 - |\Gamma|^2] = \frac{0.986}{2(63.4)}|V_0|^2$$

$$= 0.008|V_0|^2 \ W \ (98.6\% \text{ of max value})$$

For the capacitive and resistive load in series,
$$Z_L = 50 - j63.4 \ \ \Omega$$

$$\Gamma = \frac{50-j63.4-63.4}{50-j63.4+63.4} = \frac{-13.4-j63.4}{113.4-j63.4} = \frac{65∠44.5}{130∠-0.5}$$

$$= 0.5∠45.0$$

$$P_{ave} = \frac{|V_0|^2}{2Z_0}[1 - |\Gamma|^2] = \frac{1-0.25}{2(63.4)}|V_0|^2$$

$$= 0.006|V_0|^2 \ \ W \ (75\% \text{ of max value})$$

Chapter 14: Transmission Line Analysis Tools

ABCD Analysis of Transmission Lines:

There are a few popular tools to help with transmission line analyses. The first tool is the **ABCD method**. In this method, we relate the current and voltage at either end of the line with the use of matrices:

$$\begin{bmatrix} V_2 \\ I_2 \end{bmatrix} = \begin{bmatrix} A & B \\ C & D \end{bmatrix} \begin{bmatrix} V_1 \\ I_1 \end{bmatrix}$$

$$V_2 = AV_1 + BI_1; \ I_2 = CV_1 + DI_1.$$

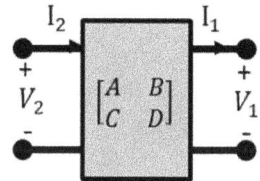

General concept for the ABCD method.

For a simple circuit, we have two cases: a parallel impedance and a series impedance:

$$\begin{bmatrix} A & B \\ C & D \end{bmatrix} = \begin{bmatrix} 1 & Z \\ 0 & 1 \end{bmatrix} \qquad \begin{bmatrix} A & B \\ C & D \end{bmatrix} = \begin{bmatrix} 1 & 0 \\ Y & 1 \end{bmatrix}$$

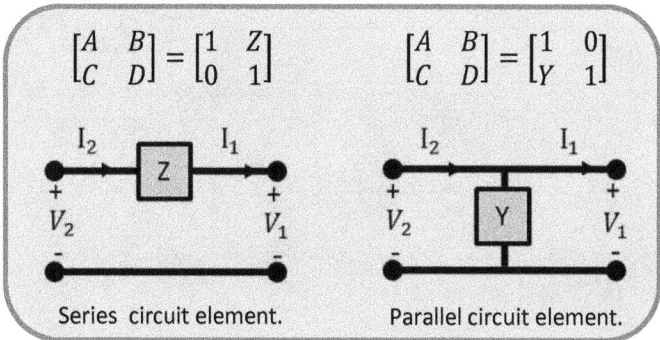

Series circuit element.   Parallel circuit element.

For a transmission line, we have:

$$\begin{bmatrix} A & B \\ C & D \end{bmatrix} = \begin{bmatrix} \cos \beta l & jZ_0 \sin \beta l \\ jY_0 \sin \beta l & \cos \beta l \end{bmatrix}.$$

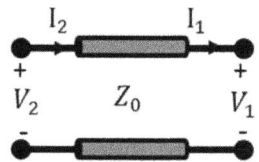

Transmission line.

Where there are multiple elements, we multiply all of the elements' matrices. The order of multiplication is important, so we work our way from V_2, I_2 to V_1, I_1:

$$\begin{bmatrix} V_2 \\ I_2 \end{bmatrix} = \begin{bmatrix} A_1 & B_1 \\ C_1 & D_1 \end{bmatrix} \begin{bmatrix} A_2 & B_2 \\ C_2 & D_2 \end{bmatrix} \cdots \begin{bmatrix} A_n & B_n \\ C_n & D_n \end{bmatrix} \begin{bmatrix} V_1 \\ I_1 \end{bmatrix}.$$

An example will illustrate how to use the ABCD method.

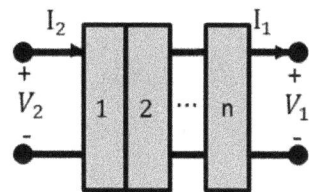

Multiple elements.

Example

Find the power delivered to the load for the following system:

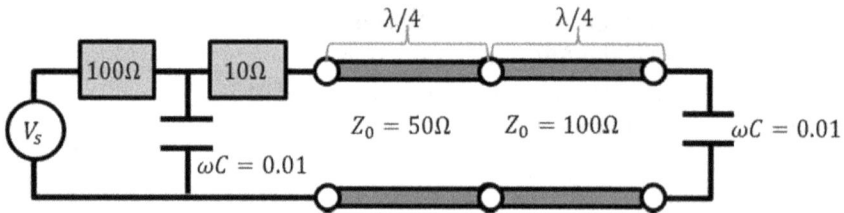

We will substitute the circuit above with an equivalent circuit as shown to the right using the ABCD method to replace all of the components except the load. The first three elements are circuit elements with:

Substitute circuit for this example.

$$Z_1 = 100; \quad Y_2 = j\omega C = j0.01; \quad Z_3 = 10.$$

The next two elements are line elements with:

$$\cos \beta l = 0; \quad \sin \beta l = 1.$$

We can now construct the ABCD relations as follows:

$$\begin{bmatrix} V_s \\ I_s \end{bmatrix} = \begin{bmatrix} 1 & 100 \\ 0 & 1 \end{bmatrix} \begin{bmatrix} 1 & 0 \\ j0.01 & 1 \end{bmatrix} \begin{bmatrix} 1 & 10 \\ 0 & 1 \end{bmatrix} \begin{bmatrix} 0 & j50 \\ j0.02 & 0 \end{bmatrix} \begin{bmatrix} 0 & j100 \\ j0.01 & 0 \end{bmatrix} \begin{bmatrix} V_L \\ I_L \end{bmatrix}$$

$$\begin{bmatrix} V_s \\ I_s \end{bmatrix} = \begin{bmatrix} 1+j & 100 \\ j0.01 & 1 \end{bmatrix} \begin{bmatrix} 1 & 10 \\ 0 & 1 \end{bmatrix} \begin{bmatrix} -0.5 & 0 \\ 0 & -2 \end{bmatrix} \begin{bmatrix} V_L \\ I_L \end{bmatrix}$$

$$\begin{bmatrix} V_s \\ I_s \end{bmatrix} = \begin{bmatrix} 1+j & 100 \\ j0.01 & 1 \end{bmatrix} \begin{bmatrix} -0.5 & -20 \\ 0 & -2 \end{bmatrix} \begin{bmatrix} V_L \\ I_L \end{bmatrix}$$

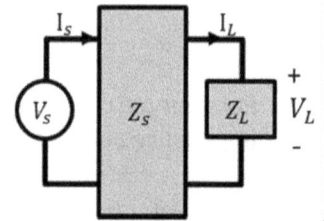

Ex. cont.

Finally, we have:

$$\begin{bmatrix} V_s \\ I_s \end{bmatrix} = \begin{bmatrix} -0.5 - j0.5 & -220 - j20 \\ -j0.005 & -2 - j0.2 \end{bmatrix} \begin{bmatrix} V_L \\ I_L \end{bmatrix}$$

Using $I_L = V_L Y_L = j0.01\, V_L$, we get:

$$V_s = V_L(-0.5 - j0.5 + 0.2 - j2.2)$$

$$= V_L(-0.3 - j2.7)$$

$$= V_L\, 2.72 \angle 4.32$$

$$V_L = V_s\, 0.368 \angle -4.32 = -1.41 + j0.34\ V_s$$

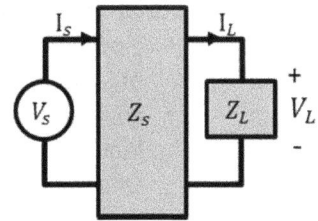

Substitute circuit for this example repeated.

Using $V_L = Z_L I_L = -j100\, I_L$

$$I_s = I_L(-0.5 - 2 - j0.2)$$

$$= I_L(-2.5 - j0.2)$$

$$= I_L\, 2.5 \angle 3.22$$

$$I_L = I_s\, 0.4 \angle -3.22 \approx -0.4\, I_s$$

The power supplied by the source $P_s = \frac{1}{2}\Re\{V_s I_s{}^*\}$.

The power delivered to the load is:

$$P_d = \frac{1}{2}\Re\{V_L I_L{}^*\} = (-1.41 \cdot -0.4)\frac{1}{2}\Re\{V_s I_s{}^*\}$$

$$= 0.56\, P_s$$

So, 56% of the power supplied makes it to the load.

Smith Chart Analysis of Transmission Lines:

The picture on the next page is a **Smith Chart**. It aids in analyzing transmission lines graphically. It looks like a huge mess, but in fact it's not that bad. Let's look at some of its key characteristics:

- The outer ring has two rows of numbers that start at the left-most edge. These are distances expressed in fractional wavelengths going:
    - clockwise (outer numbers) toward the generator or source, and
    - counter-clockwise (inner numbers) toward the load.
    - One full rotation about the circle is one λ.

- The next ring in also has two rows of numbers:
    - Angle of the reflection coefficient in degrees (outer), and
    - Angle of the transmission coefficient in degrees (inner).

- A series of circles and hyperbolas encompass the interior:
    - The circles represent the real components of the impedance. If you stay on the middle horizontal line, your impedance is purely resistive.
    - The hyperbolas represent the imaginary components of the impedance for capacitive or inductive elements. Above the horizontal they are positive, below they are negative.
    - All of the impedance elements are normalized with respect to the characteristic impedance of the line.
    - The point in the middle of the graph corresponding to $z = \dfrac{Z}{Z_o} = 1 + j0$ is where the characteristic impedance is represented.

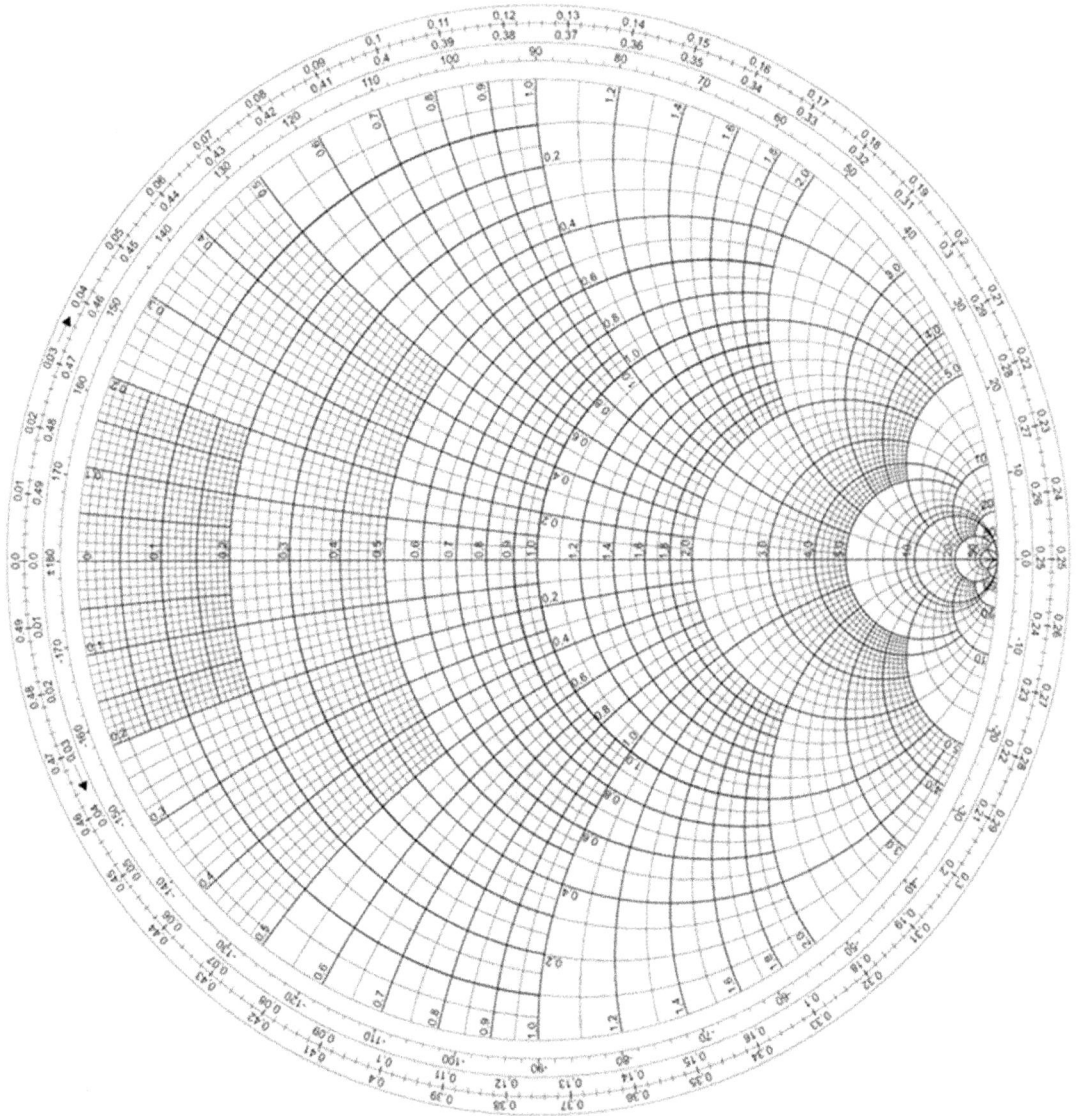

The Smith Chart.

There are a number of things that can be done with Smith Charts. We'll look at examples next.

Finding the reflectance from a load:

1. Normalize the load impedance: $z_L = \frac{Z_L}{Z_o}$ and plot it on the chart. For our example we'll use $z_L = 2 + j2$.
2. Measure the length of line $\overline{OA}$ as a percentage of the radius of the Smith chart (Point O is at $1 + j0$). For our example, we have ~0.6.
3. Extend line $\overline{OA}$ to the outer edge and read off the angle of reflection. Ours is 28°.
4. $\Gamma = |\overline{OA}| \angle refl = 0.64 \angle 28°$.

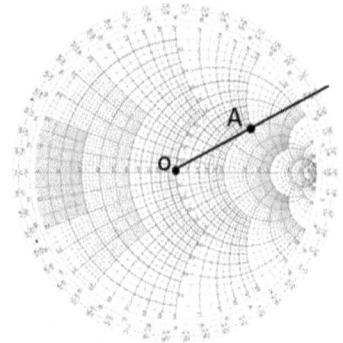

Finding the reflectance from a load.

Finding the input impedance at a distance from the load:

1. Normalize the load impedance: $z_L = \frac{Z_L}{Z_o}$ and plot it. We'll continue our example with $z_L = 2 + j2$.
2. Draw a circle with radius $|\overline{OA}|$ centered at Point O.
3. Extend line $\overline{OA}$ to reach the outer edge and read off the wavelength *toward the generator* – distances are normalize to wavelength. Our wavelength reads 0.21λ.
4. Travel clockwise around the circle the appropriate distance. For our example, let's find the input impedance at a distance 0.3λ from the load.
   - This means we want to end up at 0.51λ on the outer ring. Since the ring only goes up to 0.50λ, we'll continue around until we get to the difference, 0.01λ.
5. Draw a line from the outer edge of the circle back toward Point O. Read where this line intersects your circle at Point B. This is your normalized input impedance. Ours reads $z_{in} = 0.2 + j0.08$.
6. Multiply the answer by Z_o to get the non-normalized result.

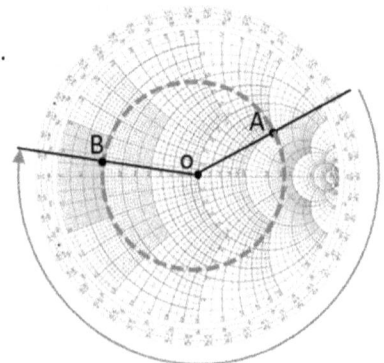

Finding the input impedance at a distance from the load.

Finding SWR and the first Vmax and Vmin from the load:

1. Plot the normalized load impedance and draw the impedance circle.
2. Read the value at Point C where the circle crosses the horizontal on the right. This is the SWR. Our is 4.0.
3. Extend line $\overline{OA}$ to the outer edge and read off the wavelength. Ours is 0.21λ.
4. The first V_{max} occurs on the horizontal line on the right. This value is 0.25λ. The distance between our point and that one is (going clockwise!) $0.25\lambda - 0.21\lambda = 0.04\lambda$.
5. The first V_{min} occurs on the horizontal line on the left. This value is 0.25λ away from the first V_{max}. The distance between our point and V_{min} is (again going clockwise!) $(0.25\lambda - 0.21\lambda) + 0.25\lambda = 0.29\lambda$.

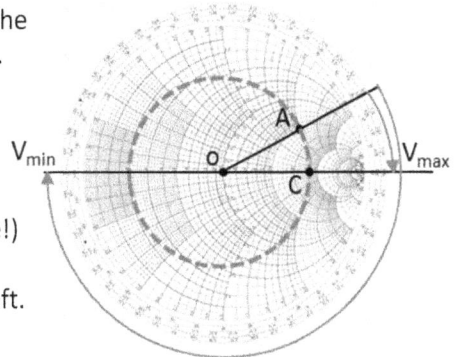

Finding the SWR, V_{max}, and V_{min} points at a distance from the load.

Designing a shorted matching stub:

1. Plot the normalized load impedance, draw the impedance circle, and extend the $\overline{OA}$ line.
2. Extend the $\overline{OA}$ line in the other direction to intersect the impedance circle. The point D is the normalized load admittance, $y_L = \frac{1}{z_L}$.
3. Find where the impedance circle crosses the $z = 1$ circle. This occurs at point E $(1 + j1.6)$ and F $(1 - j1.6)$.
4. Plot Point G at $1 - E = -j1.6$.
5. Plot Point H at $1 - F = +j1.6$.
6. Note Point J at the far right of the chart.
7. There are two solutions to the problem:
   1. The distance from the load to the first stub solution is the distance $|\overline{DE}|$. The length of the first stub solution is the distance $|\overline{JG}|$. Our first answer is a stub length of 0.09λ at a distance of 0.36λ from the load.
   2. The second solution is placed $|\overline{DF}|$ from the load with a length of $|\overline{JH}|$. Our second solution is a stub length of 0.84λ at a distance of 0.61λ.

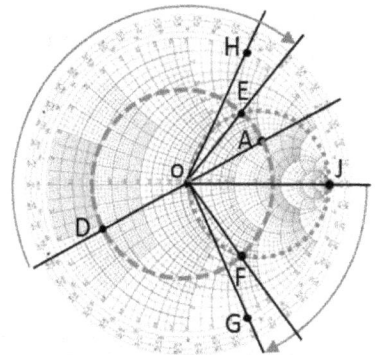

Finding the length and placement of a shorted matched stub. Two solutions are possible. The arrows indicate the first solution.

Transient Analysis of Transmission Lines:

The final analysis tool helps us figure out what happens at various points along the transmission line as reflections travel back and forth with time. Analyzing **transients** means we'll look at what happens from time $t = 0$ (when a switch closes the circuit or the generator is turned on or off) until steady state.

We will need to construct a **bounce diagram**, a plot that follows the reflections off the load and generator as they develop on the line. The horizontal axis is the length of the line, and the vertical represents time. To construct a bounce diagram, the following steps are needed:

1. Determine Γ_L and Γ_g, the reflection coefficients of the load and generator.

2. Determine I, V at $t = 0^-$ at the switch just before the change.

3. Determine I, V at $t = 0^+$ at the switch just after the change. Note that just after the change, there is no time to see the other side of the line.

4. Calculate the time, τ, for a single pass.

5. Fill in the values of I, V back and forth across the diagram.

6. Once the diagram is complete, add up the values along any vertical line on the diagram.

Once again, an example or two becomes very valuable.

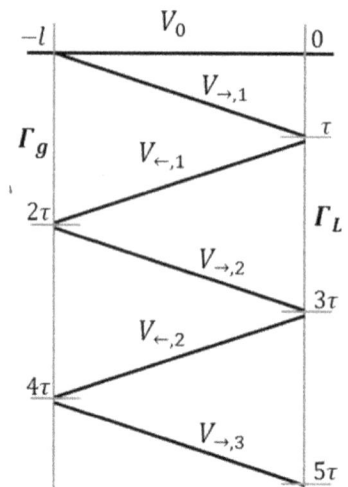

A voltage bounce diagram for analyzing transients.

Find the transient response of the line shown.

Example

$$\Gamma_L = \frac{30-10}{30+10} = \frac{1}{2}; \ \Gamma_g = \frac{5-10}{5+10} = \frac{-1}{3}$$

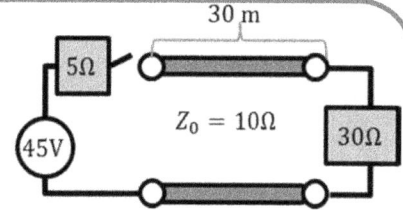

Transmission line for this example. Switch is thrown at $t = 0$.

At $t = 0^-$, $I, V = 0$ at the switch.

At $t = 0^+$, the first wave will see only the beginning of the line and its *characteristic impedance*. (It won't see the input impedance as it can't experience the load instantaneously.)

$$I(t = 0^+) = \frac{45}{15} = 3 \text{ A};$$

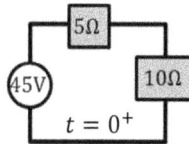

$$V(t = 0^+) = I(Z_0) = 30 \text{ V}$$

The time to traverse the line is $\tau = \frac{l}{c} = 10 \ \mu s$.

To construct the diagram for voltage, we use $V = 0$ across the top rail. The first wave will travel to the right at a value of 30 V. Upon reaching the load, it will be reflected with a value of $\Gamma_L V = 15$ V. Upon reaching the generator, it will again bounce back this time with a value of $\Gamma_g V = -5$ V, and so on.

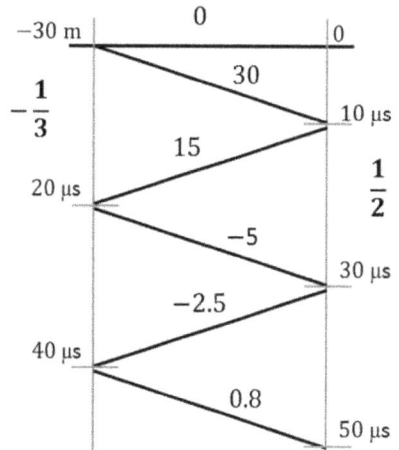

Voltage bounce diagram for this example.

Once we have enough bounces, we can then construct the time diagram. We draw a vertical line and sum every bar we cross with time:

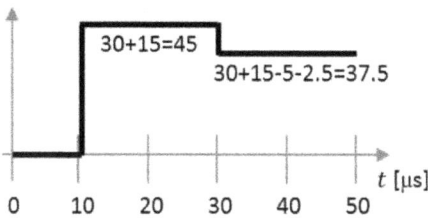

Voltage transient response at load.

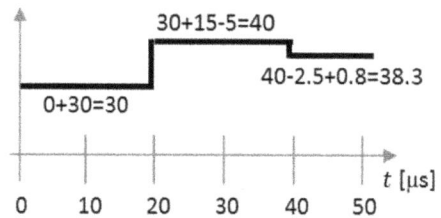

Voltage transient response at switch.

Ex. cont.

The current diagram will look similar, except remember that the reflection coefficients for current are the negative of those for voltage.

Along with drawing the current transient response at the generator and the load, the transients on the mid-point of the line (dotted grey line on the bounce diagram) is also shown.

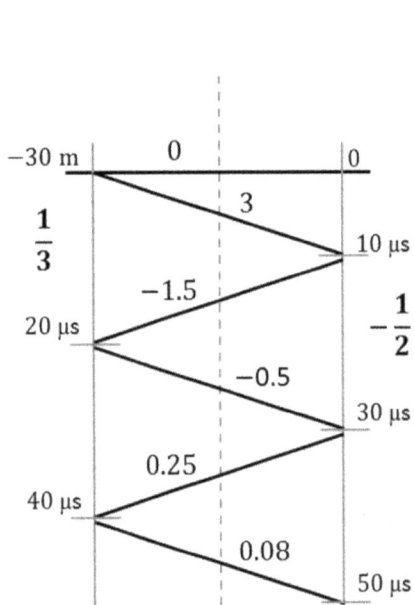

I at 0 m

3-1.5=1.5

1.5-0.5+0.25=1.25

t [μs]

0   10   20   30   40   50

−30 m   0   0

$\frac{1}{3}$ 3

10 μs

−1.5

20 μs $-\frac{1}{2}$

−0.5

30 μs

I at -15 m

3

1.5

1   1.25 1.33

t [μs]

0   10   20   30   40   50

0.25

40 μs

0.08

50 μs

Current bounce diagram for this example.

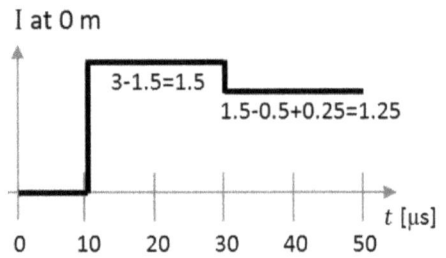

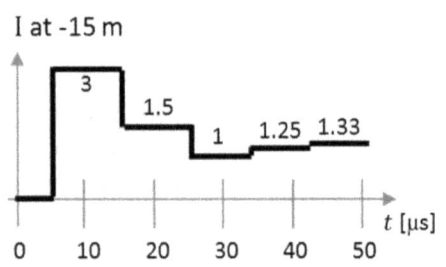

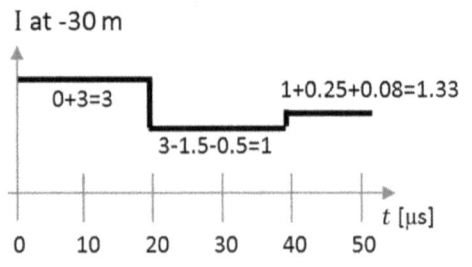

I at -30 m

0+3=3

1+0.25+0.08=1.33

3-1.5-0.5=1

t [μs]

0   10   20   30   40   50

Current transient response.

Find the voltage transient response of the prior example with the switch placed in the middle of the transmission line.

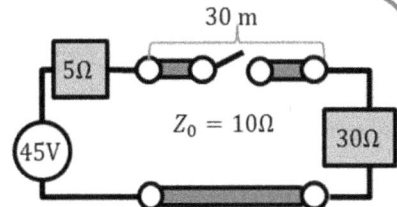

30 m

$Z_0 = 10\Omega$

Transmission line for this example. Switch is thrown at $t = 0$.

We start the same as before:

$$\Gamma_L = \frac{30-10}{30+10} = \frac{1}{2}; \ \Gamma_g = \frac{5-10}{5+10} = \frac{-1}{3}$$

At $t = 0^-$, we consider the case before and after the switch:

	$z < 15$ m	$z > 15$ m
V	45	0
I	0	0

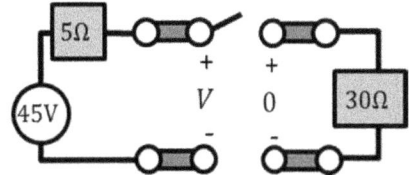

Situation at $t = 0^-$.

At $t = 0^+$, the first wave will see only the line right at the switch. Current and voltage will begin to flow in both directions. The 45 V will disperse, half forward and half backward so that a total of 22.5 V is present on both sides of the line. Current will flow as $I_\rightarrow = {}^{22.5}/_{10} = 2.25$ A and $I_\leftarrow = -{}^{-22.5}/_{10} = 2.25$ A:

	$\leftarrow$	$\rightarrow$
V	-22.5	22.5
I	2.5	2.5

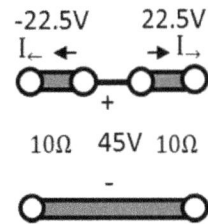

Situation at $t = 0^+$.

(note $I_\rightarrow = I_\leftarrow$ as per Kirchhoff's current law.)

The time to traverse the line is still $\tau = \frac{l}{c} = 10 \ \mu s$.

We can draw the voltage bounce diagram using the right starting conditions and propagating two waves, forward (shown in black) and backward (shown in grey). The current transient response can be found in a similar fashion. When adding up the bars for the timing diagram, don't forget the top rail.

Note that the answer is eventually converging to the steady-state response just as in the prior example where:

$$I_{ss} = \frac{V_g}{Z_{total}} = \frac{45}{5+35} = 1.29 \text{ A};$$

$$V_{L,ss} = I_{ss}Z_L = 38.6 \text{ V}.$$

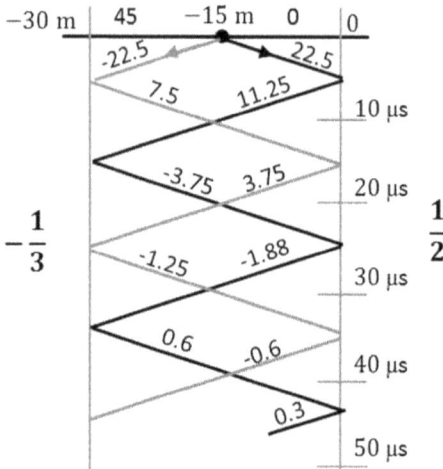

V at 0 m

Voltage bounce diagram for this example.

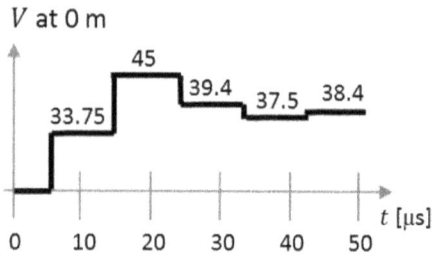

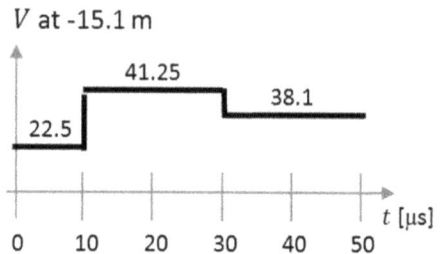

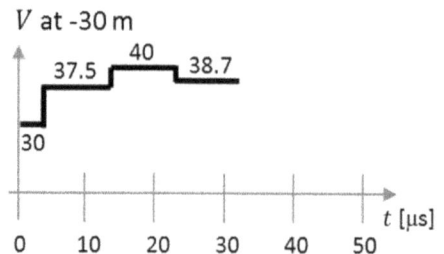

V at -15.1 m

V at -30 m

Voltage transient response at various points on the line.

Appendices

A: Course Summary
B: Vector Calculus Review
C: Units & Constants

Appendix A: Course Summary

Statics Summary

	Electro-Statics	Magneto-Statics
Sources	$Q = \sum q_i + \int \rho_L dL + \cdots$	$I = \frac{-dQ}{dt} = \int \vec{J} \cdot d\vec{S}$
Material	$\vec{D} = \varepsilon\vec{E}; \quad \vec{J} = \sigma\vec{E}$ $\vec{P} = \varepsilon_o \chi_e \vec{E}$ $\varepsilon = \varepsilon_o(1 + \chi_e) = \varepsilon_0 \varepsilon_r$	$\vec{B} = \mu\vec{H}$ $\vec{M} = \mu_0 \chi_m \vec{H}$ $\mu = \mu_o(1 + \chi_m) = \mu_0 \mu_r$
M.E. Point	$\nabla \times \vec{E} = 0; \qquad \nabla \cdot \vec{D} = \rho_v$	$\nabla \times \vec{H} = \vec{J}; \qquad \nabla \cdot \vec{B} = 0$
M.E. Integ.	$\oint \vec{E} \cdot d\vec{L} = 0; \qquad \oiint \vec{D} \cdot d\vec{S} = Q$	$\oint \vec{H} \cdot dL = I; \qquad \oiint \vec{B} \cdot d\vec{S} = 0$
Fields & Sources	$\vec{E} = \sum \frac{q_i}{4\pi\varepsilon r_o^2} \hat{r}_o + \frac{1}{4\pi\varepsilon} \int \frac{\rho_L}{r_o^2} \hat{r}_o \, dL' + \cdots$	$\vec{H} = \frac{1}{4\pi} \int \frac{I\,d\vec{L'} \times \hat{r}_o}{r_o^2} = \frac{1}{4\pi} \int \frac{\vec{J} \times \hat{r}_o}{r_o^2} dv'$
Potential	$V_{ab} = V_b - V_a = -\int_a^b \vec{E} \cdot d\vec{L}$ $V = \sum \frac{q_i}{4\pi\varepsilon r_o} + \frac{1}{4\pi\varepsilon} \int \frac{\rho_L}{r_o} dL' + \cdots$ $\nabla^2 V = \frac{-\rho_v}{\varepsilon}; \quad \vec{E} = -\nabla V$	$\vec{A} = \frac{\mu}{4\pi} \int \frac{I\,d\vec{L'}}{r_o} = \frac{\mu}{4\pi} \int \frac{\vec{J}}{r_o} dv'$ $\nabla^2 \vec{A} = -\mu\vec{J}; \quad \vec{B} = \nabla \times \vec{A}$
Boundary Conds.	$D_{n1} - D_{n2} = \rho_s$ $\varepsilon_1 E_{n1} - \varepsilon_2 E_{n2} = \rho_s$ $\varepsilon_1 D_{t1} = \varepsilon_2 D_{t2}$ $E_{t1} = E_{t2}$	$B_{n1} = B_{n2}$ $\mu_1 H_{n1} = \mu_2 H_{n2}$ $\mu_2 B_{t1} = \mu_1 B_{t2}$ $H_{t1} = H_{t2}$
Dipoles & Moments	$\vec{E}_{dip} = \frac{qd}{4\pi\varepsilon r^3}[2\cos\theta\,\hat{r} + \sin\theta\,\hat{\theta}]$ $\vec{p} = q\vec{d}$	$\vec{H}_{dip} = \frac{m}{4\pi r^3}[2\cos\theta\,\hat{r} + \sin\theta\,\hat{\theta}]$ $\vec{m} = I\pi a^2 \vec{z}$
Circuit Elements	$R = \frac{V}{I} = \frac{-\int \vec{E} \cdot d\vec{L}}{\int \sigma\vec{E} \cdot d\vec{S}}$ $C = \frac{Q}{V} = \frac{\int \varepsilon\vec{E} \cdot d\vec{S}}{-\int \vec{E} \cdot d\vec{L}}$	$L = \frac{N\Phi}{I} = \frac{N}{I} \int \vec{B} \cdot d\vec{S}$ $L_{12} = \frac{N_2}{I_1} \int \vec{B_1} \cdot d\vec{S_2}$
Energy & Power	$W_e = \int \varepsilon |\vec{E}|^2 \, dv$ $P = \int \vec{E} \cdot \vec{J} \, dv = \int \sigma |\vec{E}|^2 \, dv$	$W_m = \int \mu |\vec{H}|^2 \, dv$
Forces & Torque	$\vec{F_e} = q'\vec{E} = -\nabla W_e$	$\vec{F_m} = q'\vec{v} \times \vec{B} = I \int d\vec{L} \times \vec{B} = -\nabla W_m$ $\vec{T} = \vec{d} \times \vec{F} = \vec{m} \times \vec{B}$

Dynamics Summary

	Electric (new or changed)	Magnetic (new or changed)
Sources & Forces	$V_{emf} = \frac{-N\Phi}{dt} = -N\frac{d}{dt}\int \vec{B} \cdot d\vec{S}$ $= \oint \vec{E} \cdot d\vec{L}$	$I = \frac{-dQ}{dt} = \int \left(\vec{J} + \frac{d\vec{D}}{dt}\right) \cdot d\vec{S}$; $\nabla \cdot \vec{J} = \frac{-d\rho_v}{dt}$
M.E. Point	$\nabla \times \vec{E} = -\frac{d\vec{B}}{dt}$; $\quad \nabla \cdot \vec{D} = \rho_v$; $\nabla^2 \vec{E} + k^2 \vec{E} = 0$	$\nabla \times \vec{H} = \vec{J} + \frac{d\vec{D}}{dt}$; $\quad \nabla \cdot \vec{B} = 0$; $\nabla^2 \vec{H} + k^2 \vec{H} = 0$
M.E. Integ.	$\oint \vec{E} \cdot d\vec{L} = \frac{-d}{dt}\int \vec{B} \cdot d\vec{S}$; $\oiint \vec{D} \cdot d\vec{S} = Q$	$\oint \vec{H} \cdot d\vec{L} = \int (\vec{J} + \frac{d\vec{D}}{dt}) \cdot d\vec{S}$; $\oiint \vec{B} \cdot d\vec{S} = 0$
Potential	$\nabla^2 V = \frac{-\rho_v}{\varepsilon} + \mu\varepsilon \frac{d^2 V}{dt^2}$; $\vec{E} = -\nabla V - \frac{d\vec{A}}{dt}$	$\nabla^2 \vec{A} = -\mu\vec{J} + \mu\varepsilon \frac{d^2\vec{A}}{dt^2}$; $\nabla \cdot \vec{A} = -\mu\varepsilon \frac{dV}{dt}$
Energy & Power Density	$W_e = \frac{1}{2}\int \varepsilon \left|\vec{E}\right|^2 dv$ $\vec{\mathcal{P}} = \vec{E} \times \vec{H}$	$W_m = \frac{1}{2}\int \mu \left|\vec{H}\right|^2 dv$ $\vec{\mathcal{P}}_{ave} = \frac{1}{2} Re\{\vec{E} \times \vec{H}^*\}$

Boundary Equations	$\Gamma = \frac{\left|\vec{E}_r\right|}{\left|\vec{E}_i\right|} = \frac{\eta_2 - \eta_1}{\eta_2 + \eta_1}$	$\tau = \frac{2\eta_2}{\eta_2 + \eta_1} = 1 + \Gamma$	$SWR = \frac{1 + |\Gamma|}{1 - |\Gamma|}$
Replace-ment η:	$\eta_2 \equiv \eta_{in} = \frac{\eta_3 \cos k_2 l + j\eta_2 \sin k_2 l}{\eta_2 \cos k_2 l + j\eta_3 \sin k_2 l}$	$\eta_i \equiv \eta'_i = \eta_i \sec\theta_i$	$\eta_i \equiv \eta'_i = \eta_i \cos\theta_i$
Condition:	Multiple boundaries	s-polarization, obl.	p-polarization, obl.

Reflection	Refraction	Critical Angle	Brewster's Angle
$\theta_i = \theta_r \equiv \theta_1$	$n_1 \sin\theta_1 = n_2 \sin\theta_2$	$\sin\theta_c = \frac{n_2}{n_1}$; $n_1 > n_2$	$\sin\theta_B = n_2 \sqrt{\frac{1}{n_1^2 + n_2^2}}$; p-polarization

Dielectric	$\alpha \approx 0$	$\beta = k \approx \omega\sqrt{\mu\varepsilon'}$	$\eta = \frac{\left|\vec{E}\right|}{\left|\vec{H}\right|} \approx \sqrt{\frac{\mu}{\varepsilon'}}$
Conductor	$\alpha = \beta \approx \sqrt{\frac{\sigma\omega\mu}{2}}$	$k \approx \sqrt{\frac{\sigma\omega\mu}{2}}(1 - j)$	$\eta = \frac{\left|\vec{E}\right|}{\left|\vec{H}\right|} \approx \sqrt{\frac{\omega\mu}{2\sigma}}(1 + j)$

Transmission Line Summary

General

Transmission Line Equations	$\dfrac{dV}{dz} = -(R + j\omega L)\mathrm{I}$ $\dfrac{d\mathrm{I}}{dz} = -(G + j\omega C)V$
Propagation Constant	$\gamma^2 = (R + j\omega L)(G + j\omega C)$
Characteristic Impedance	$Z_0 = \dfrac{V_\rightarrow}{\mathrm{I}_\rightarrow} = \sqrt{\dfrac{R + j\omega L}{G + j\omega C}} = -\dfrac{V_\leftarrow}{\mathrm{I}_\leftarrow}$
Input Impedance	$Z_{in}(z) = \dfrac{V(z)}{\mathrm{I}(z)} = Z_0 \dfrac{1 + \Gamma e^{j2\beta z}}{1 - \Gamma e^{j2\beta z}}$ $Z_{in} = Z_0 \dfrac{Z_L + jZ_0 \tan \beta l}{Z_0 + jZ_L \tan \beta l}$
Reflectance Coefficient	$\Gamma = \dfrac{V_\leftarrow}{V_\rightarrow} = \dfrac{Z_L - Z_0}{Z_L + Z_0} = -\dfrac{\mathrm{I}_\leftarrow}{\mathrm{I}_\rightarrow}$
Standing Wave Ratio	$SWR = \dfrac{|V_{max}|}{|V_{min}|} = \dfrac{1 + |\Gamma|}{1 - |\Gamma|}$
Time-Averaged Power Delivered	$P_{ave} = \dfrac{|V_0|^2}{2Z_0}[1 - |\Gamma|^2] =$ $\dfrac{1}{2}\mathfrak{Re}\{\vec{V} \cdot \vec{\mathrm{I}}^*\}$

ABCD

Circuit	Parallel Z	$\begin{bmatrix} 1 & Z \\ 0 & 1 \end{bmatrix}$
	Series Z	$\begin{bmatrix} 1 & 0 \\ Y & 1 \end{bmatrix}$
	Transmission Line	$\begin{bmatrix} \cos \beta l & jZ_0 \sin \beta l \\ jY_0 \sin \beta l & \cos \beta l \end{bmatrix}$

Transient

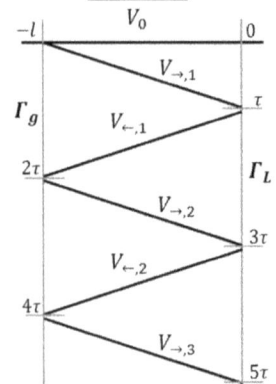

Find Γ_L, Γ_g, τ and V, I at $t = 0^-, 0^+$
For $\mathrm{I}, \Gamma_i = -\Gamma_i$
Each pass is V (or I) $* \Gamma_i$
Draw vertical and add all crossing values.

Smith Chart

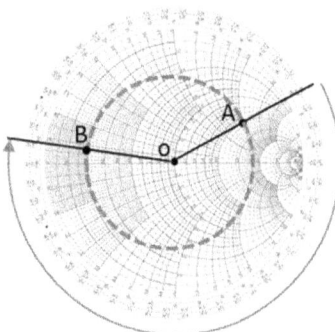

$A = z_L = Z_L/Z_0$
$\Gamma = |\overline{OA}|\angle refl$
(angle on outer edge)
$B = Z_{in}$ at $|\overline{AB}|$ from load

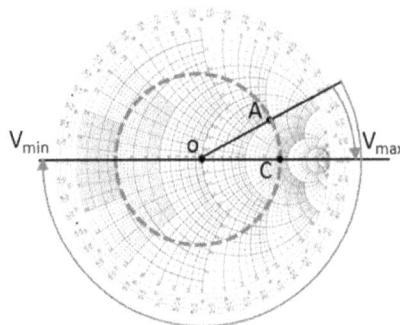

$C = SWR$
1st V_{max} is at $|\overline{AC}|$ from load
1st V_{min} is at $|\overline{AC}| + \lambda/2$

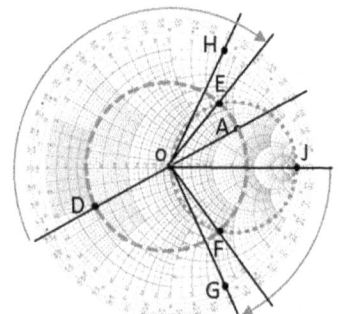

$A = z_L$; $D = y_L$
E & F $= \cap [\overline{OA} \odot, z = 1]$
$G = 1 - $ E & H $= 1 - $ F
Stub 1: $l = |\overline{JG}|$; $d = |\overline{DE}|$ (shown)
Stub 2: $l = |\overline{JH}|$; $d = |\overline{DF}|$

Appendix B: Vector Calculus Review

Rectangular

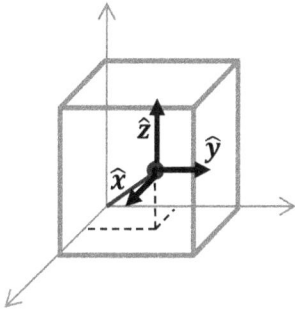

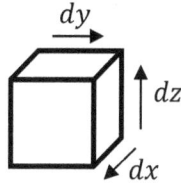

$$d\vec{L} = dx\,\hat{x} + dy\,\hat{y} + dz\,\hat{z}$$

$$d\hat{S} = dx\,dy\,\hat{z}$$

$$d\hat{S} = dx\,dz\,\hat{y}$$

$$d\hat{S} = dy\,dz\,\hat{x}$$

$$dv = dx\,dy\,dz$$

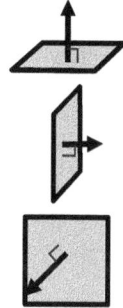

Cylindrical

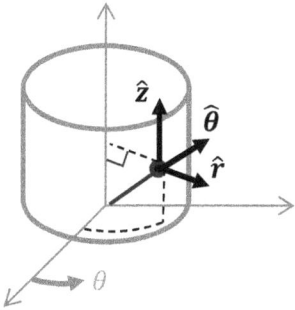

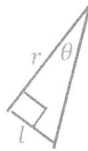

$dl = rd(\sin\theta)\sim rd\theta$

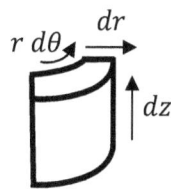

$$d\vec{L} = dr\,\hat{r} + rd\theta\,\hat{\theta} + dz\,\hat{z}$$

$$d\hat{S} = r\,dr\,d\theta\,\hat{z}$$

$$d\hat{S} = r\,d\theta\,dz\,\hat{r}$$

$$d\hat{S} = dr\,dz\,\hat{\theta}$$

$$dv = r\,dr\,d\theta\,dz$$

Spherical

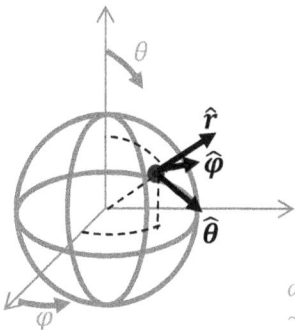

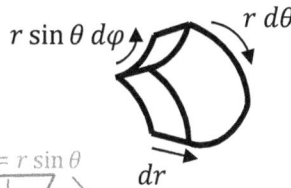

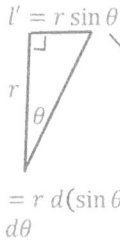

$l' = r\sin\theta$

$dl' = r\,d(\sin\theta)$
$\sim r\,d\theta$

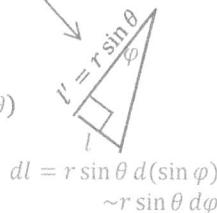

$dl = r\sin\theta\,d(\sin\varphi)$
$\sim r\sin\theta\,d\varphi$

$$d\vec{L} = dr\,\hat{r} + r\,d\theta\,\hat{\theta} + r\sin\theta\,d\varphi\,\hat{\varphi}$$

$$d\hat{S} = r^2\sin\theta\,d\theta\,d\varphi\,\hat{r}$$

$$dv = r^2\sin\theta\,dr\,d\theta\,d\varphi$$

Remember: To integrate, use $\theta\epsilon[0,\pi]$, $\varphi\epsilon[0,2\pi]$ to avoid covering the same volume twice

Rectangular

$$\nabla V = \frac{\partial V}{\partial x}\,\hat{x} + \frac{\partial V}{\partial y}\,\hat{y} + \frac{\partial V}{\partial z}\,\hat{z}$$

$$\nabla \cdot \vec{A} = \frac{\partial A_x}{\partial x} + \frac{\partial A_y}{\partial y} + \frac{\partial A_z}{\partial z}$$

$$\nabla \times \vec{A} = \begin{vmatrix} \hat{x} & \hat{y} & \hat{z} \\ \frac{\partial}{\partial x} & \frac{\partial}{\partial y} & \frac{\partial}{\partial z} \\ A_x & A_y & A_z \end{vmatrix} = \left(\frac{\partial A_z}{\partial y} - \frac{\partial A_y}{\partial z}\right)\hat{x} + \left(\frac{\partial A_x}{\partial z} - \frac{\partial A_z}{\partial x}\right)\hat{y} + \left(\frac{\partial A_y}{\partial x} - \frac{\partial A_x}{\partial y}\right)\hat{z}$$

$$\nabla^2 V = \frac{\partial^2 V}{\partial x^2} + \frac{\partial^2 V}{\partial y^2} + \frac{\partial^2 V}{\partial z^2}$$

Cylindrical

$$\nabla V = \frac{\partial V}{\partial r}\,\hat{r} + \frac{1}{r}\frac{\partial V}{\partial \theta}\,\hat{\theta} + \frac{\partial V}{\partial z}\,\hat{z}$$

$$\nabla \cdot \vec{A} = \frac{1}{r}\frac{\partial (r\,A_r)}{\partial r} + \frac{1}{r}\frac{\partial A_\theta}{\partial \theta} + \frac{\partial A_z}{\partial z}$$

$$\nabla \times \vec{A} = \frac{1}{r}\begin{vmatrix} \hat{r} & r\hat{\theta} & \hat{z} \\ \frac{\partial}{\partial r} & \frac{\partial}{\partial \theta} & \frac{\partial}{\partial z} \\ A_r & rA_\theta & A_z \end{vmatrix} = \left(\frac{1}{r}\frac{\partial A_z}{\partial \theta} - \frac{\partial A_\theta}{\partial z}\right)\hat{r} + \left(\frac{\partial A_r}{\partial z} - \frac{\partial A_z}{\partial r}\right)\hat{\theta} + \frac{1}{r}\left(\frac{\partial (rA_\theta)}{\partial r} - \frac{\partial A_r}{\partial \theta}\right)\hat{z}$$

$$\nabla^2 V = \frac{1}{r}\frac{\partial}{\partial r}\left(r\frac{\partial V}{\partial r}\right) + \frac{1}{r^2}\frac{\partial^2 V}{\partial \theta^2} + \frac{\partial^2 V}{\partial z^2}$$

Spherical

$$\nabla V = \frac{\partial V}{\partial r}\,\hat{r} + \frac{1}{r}\frac{\partial V}{\partial \theta}\,\hat{\theta} + \frac{1}{r\sin\theta}\frac{\partial V}{\partial \varphi}\,\hat{\varphi}$$

$$\nabla \cdot \vec{A} = \frac{1}{r^2}\frac{\partial (r^2 A_r)}{\partial r}\,\hat{r} + \frac{1}{r\sin\theta}\frac{\partial (A_\theta \sin\theta)}{\partial \theta}\,\hat{\theta} + \frac{1}{r\sin\theta}\frac{\partial A_\varphi}{\partial \varphi}\,\hat{\varphi}$$

$$\nabla \times \vec{A} = \frac{1}{r^2 \sin\theta}\begin{vmatrix} \hat{r} & r\hat{\theta} & r\sin\theta\,\hat{\varphi} \\ \frac{\partial}{\partial r} & \frac{\partial}{\partial \theta} & \frac{\partial}{\partial \varphi} \\ A_r & rA_\theta & r\sin\theta\,A_\varphi \end{vmatrix}$$

$$= \frac{1}{r\sin\theta}\left(\frac{\partial (A_\varphi \sin\theta)}{\partial \theta} - \frac{\partial A_\theta}{\partial \varphi}\right)\hat{r} + \frac{1}{r}\left(\frac{1}{\sin\theta}\frac{\partial A_r}{\partial \varphi} - \frac{\partial (rA_\varphi)}{\partial r}\right)\hat{\theta} + \frac{1}{r}\left(\frac{\partial (rA_\theta)}{\partial r} - \frac{\partial A_r}{\partial \theta}\right)\hat{\varphi}$$

$$\nabla^2 V = \frac{1}{r^2}\frac{\partial}{\partial r}\left(r^2 \frac{\partial V}{\partial r}\right) + \frac{1}{r^2 \sin\theta}\frac{\partial}{\partial \theta}\left(\sin\theta \frac{\partial V}{\partial \theta}\right) + \frac{1}{r^2 \sin^2\theta}\frac{\partial^2 V}{\partial \varphi^2}$$

	Convert to Cylindrical	**Convert to Spherical**
Rectangular	$\hat{x} \times \hat{y} = \hat{z}$ $\hat{y} \times \hat{z} = \hat{x}$ $\hat{z} \times \hat{x} = \hat{y}$	$r = \sqrt{x^2 + y^2}$ $\theta = \tan^{-1}\left(\frac{y}{x}\right)$ $z = z$ $\hat{r} = \hat{x}\cos\theta + \hat{y}\sin\theta$ $\hat{\theta} = -\hat{x}\sin\theta + \hat{y}\cos\theta$ $\hat{z} = \hat{z}$	$r = \sqrt{x^2 + y^2 + z^2}$ $\theta = \tan^{-1}\left(\frac{\sqrt{x^2+y^2}}{z}\right)$ $\varphi = \tan^{-1}\left(\frac{y}{x}\right)$ $\hat{r} = \hat{x}\sin\theta\cos\varphi + \hat{y}\sin\theta\sin\varphi + \hat{z}\cos\theta$ $\hat{\theta} = \hat{x}\cos\theta\cos\varphi + \hat{y}\cos\theta\sin\varphi - \hat{z}\sin\theta$ $\hat{\varphi} = -\hat{x}\sin\varphi + \hat{y}\cos\varphi$

	Convert to Cartesian	**Convert to Spherical**
Cylindrical	$\hat{r} \times \hat{\theta} = \hat{z}$ $\hat{\theta} \times \hat{z} = \hat{r}$ $\hat{z} \times \hat{r} = \hat{\theta}$	$x = r\cos\theta$ $y = r\sin\theta$ $z = z$ $\hat{x} = \hat{r}\cos\theta - \hat{\theta}\sin\theta$ $\hat{y} = \hat{r}\sin\theta + \hat{\theta}\cos\theta$ $\hat{z} = \hat{z}$	$r_s = \sqrt{r_c^2 + z^2}$ $\theta = \tan^{-1}\left(\frac{r_c}{z}\right)$ $\varphi = \theta \text{ (cyl)}$ $\hat{r}_s = \hat{r}_c \sin\left[\tan^{-1}\left(\frac{r_c}{z}\right)\right] + \hat{z}\cos\left[\tan^{-1}\left(\frac{r_c}{z}\right)\right]$ $\hat{\theta} = \hat{r}_c \cos\left[\tan^{-1}\left(\frac{r_c}{z}\right)\right] - \hat{z}\sin\left[\tan^{-1}\left(\frac{r_c}{z}\right)\right]$ $\hat{\varphi} = \hat{\theta}$

	Convert to Cylindrical	**Convert to Cartesian**
Spherical	$\hat{r} \times \hat{\theta} = \hat{\varphi}$ $\hat{\theta} \times \hat{\varphi} = \hat{r}$ $\hat{\varphi} \times \hat{r} = \hat{\theta}$	$r_c = r_s \sin\theta$ $\theta = \varphi$ $z = r_s \cos\theta$ $\hat{r}_c = \hat{r}_s \sin\theta + \hat{\theta}\cos\theta$ $\hat{\theta} = \hat{\varphi}$ $\hat{z} = \hat{r}_c \cos\theta - \hat{\theta}\sin\theta$	$x = r\sin\theta\cos\varphi$ $y = r\sin\theta\sin\varphi$ $z = r\cos\theta$ $\hat{x} = \hat{r}\sin\theta\cos\varphi + \hat{\theta}\cos\theta\cos\varphi - \hat{\varphi}\sin\varphi$ $\hat{y} = \hat{r}\sin\theta\sin\varphi + \hat{\theta}\cos\theta\sin\varphi + \hat{\varphi}\cos\varphi$ $\hat{z} = \hat{r}\cos\theta - \hat{\theta}\sin\theta$

Note: From cylindrical to spherical coordinates, I switch θ from measuring off the $\hat{x}$ axis to measuring off the $\hat{z}$ axis so that integrals with respect to θ always use "$r\partial\theta$" in the integrand no matter the coordinate system.

Appendix C: Units & Constants

Element	Symb.	Units	Element	Symb.	Units
Electric Field	$\mathbf{E}$	V/m	Conductance	σ	S/m
Electric Flux Density	$\mathbf{D}$	C/m^2	Resistance	R	Ω
Permittivity	ε	F/m	Capacitance	C	F
Magnetic Field	$\mathbf{H}$	A/m	Flux, Flux Linkage	Φ, Λ	Wb
Magnetic Flux Density	$\mathbf{B}$	T	Inductance	L	Wb/A
Permeability	μ	H/m	Power	P	P
Point Charge	q	C	Electric Energy	W_e	J
Total Charge	Q	C	Magnetic Energy	W_m	J
Line Charge Density	ρ_L	C/m	Force	F	N
Surface Charge Density	ρ_S	C/m^2	Torque	T	N·m
Volume Charge Density	ρ_V	C/m^3	Frequency	ω	rad/s
Current	I	A	Frequency	f	Hz
Current Density	$\mathbf{J}$	A/m^2	Wavelength	λ	m
Potential, Voltage, emf	V	V	Phase	ϕ	rad
Magnetic Potential	$\mathbf{A}$	Wb/m	Phase Velocity	v_p	m/s
Electric Dipole Moment	$\mathbf{p}$	C·m	Wavenumber	k	1/m
Magnetic Dipole Moment	$\mathbf{m}$	$A·m^2$	Impedance	η, Z	Ω

Free Space Permittivity	ε_o	8.854×10^{-12}	F/m
Free Space Permeability	μ_o	$4\pi \times 10^{-7}$	H/m
Free Space Intrinsic Impedance	η_o	$120\pi = 377$	Ω
Speed of Light	c	3×10^8	m/s

Index

A-B

ABCD method	109
Ampère's law	7, 8, 64
angular momentum	52
attenuation constant	89-90
Biot-Savart law	30
bounce diagram	116
boundary conditions	
for dyanamics	66
normal boundary conditions	37
tangential boundary conditions	38
Brewster's angle	100

C

capacitance	43
characteristic impedance	102
charge sources	
infinite line charge	15
infinite sheet charge	15
line charge density	13
surface charge density	13
volume charge	7
volume charge density	13
charge continuity equation	64
circular polarization	80
complex propagation constant	102
conductance	27, 43

charge continuity equation	64
circular polarization	80
complex propagation constant	102
conductance	27, 43
consitutent relations	5, 6, 66
Coulomb force	53
Coulomb's law	17, 53
critical angle	99
current	
current density	7, 13, 27
displacement current	64
displacement current density	64

D-E

dipole	
electric dipole	25
electric dipole moment	25, 51
magnetic dipole	29, 35
magnetic dipole moment	35, 51
electric dipole	25
electric dipole moment	25, 51
electric field	
electric field intensity	5
electric flux density	5
electric force	53
electric potential	18, 67-68
electric susceptibility	16

electro-magnetic (EM) wave 71,74

electro-magnetic induction 61

electro-motive force 61

   general emf 61

   motional emf 61

   transformer emf 61

electro-static potential energy 48

elliptical polarization 80

equipotential surface 41

F-H

Faraday's law 61

force

   Coulomb force 53

   electric force 53

   electro-motive force (emf) 61

   Lorentz force 54

   magnetic force 54

frequency 73

Gauss's laws

   Gauss's law for electro-statics 7, 8

   Gauss's law for magneto-statics 7, 8

general emf 61

Helmholtz equation 75

I-K

image theory 41

impedance

   characteristic impedance 102

   input impedance 87, 105

   intrinsic impedance 74

   load impedance 103

index of refraction 73

inductance

   mutual inductance 45

   self-inductance 45

infinite line charge 15

infinite sheet charge 15

input impedance 87, 105

intrinsic impedance 74

Joule's law 48

Kirchhoff's laws 7, 8, 64

L

Laplace's equation 23

Lentz's law 61

line charge density 13

linear polarization 79

linear superposition 17

load admittance   103
load impedance   103
Lorentz force   54
lossless transmission line   101
lumped element model   101

M-O

magnetic dipole   29, 35
magnetic dipole moment   35, 51
magnetic field
   magnetization field   29
   magnetic field intensity   5, 6
   magnetic flux density   5, 6
magnetic flux   45
magnetic flux linkage   45
magnetic force   54
magnetic potential   33, 67-68
magnetic susceptibility   29
magneto-static potential energy   48
matched line   103
Maxwell's equations
   derivative form for dynamics   66
   derivative form for statics   7
   for waves   74
   history of   6
   integral form for dynamics   66
   integral form for statics   8
   point form for dynamics   66
   point form for statics   7

moments
   electric dipole moment   25, 51
   magnetic dipole moment   35, 51
   orbital magnetic moment   52
   spin magnetic moment   52
moment arm   56
motional emf   61
mutual inductance   45

normal boundary conditions   37

Ohm's law   27
orbital magnetic moment   52

P

parallel polarization   93, 96
permeability
   general   5, 6, 29
   permeability of free space   29
   relative permeability   29
permittivity
   general   5, 16
   permittivity of free space   16
   relative permittivity   16
perpendicular polariztion   93, 94-95
phase   73
phase matching condition   94
phase velocity   73
plane of incidence   85

plane wave	71
point charge	13
Poisson's equation	23
Poisson's vector equation	33
polarization	79-80, 93
circular polarization	80
elliptical polarization	80
linear polarization	79
parallel polarization	93, 96
perpendicular polarization	93, 94-95
p-polarization	93, 96
s-polarization	93, 94-95
polarizer	100
potential	
electric potential	18, 67-68
magnetic potential	33, 67-68
retarded potential	68
potential energy	
electro-static potential energy	48
magneto-static potential energy	48
potential energy in dynamics	76
power	48, 76
Poynting theorem	76
Poynting vector	76
p-polarization	93, 96
propagation	
complex propagation constant	102
constant	73, 89-90
constant of free space	73
velocity	68

Q-R

reflection coefficient	85, 103
refractive index	73
relative permeability	29
relative permittivity	16
resistance	43
retarded potential	68

S

self-inductance	45
shunt stub	106, 115
skin depth	91
Smith Chart	112-113
Snell's law of reflection	94
Snell's law of refraction	94
solenoid	32
spin magnetic moment	52
s-polarization	93, 94-95
standing wave ratio (SWR)	85, 103
susceptibility	
electric susceptibility	16
magnetic susceptibility	29
surface charge density	13

T-Z

tangential boundary conditions	38
toroid	46
torque	56
total charge	13
total internal reflection (TIR)	99
transformer	62, 105
transformer emf	61
transients	116
transmission coefficient	85
transmission line	101
transmission line equations	101
voltage	18
volume charge	7
volume charge density	13
Wave equation	75
wave number	73
wavelength	73

www.ingramcontent.com/pod-product-compliance
Lightning Source LLC
Chambersburg PA
CBHW062026210326
41519CB00060B/7181